The Insider's Guide to Arm Cortex-M Development

Leverage embedded software development tools and examples to become an efficient Cortex-M developer

Zachary Lasiuk

Pareena Verma

Jason Andrews

BIRMINGHAM—MUMBAI

The Insider's Guide to Arm Cortex-M Development

Group Product Manager: Rahul Nair
Publishing Product Manager: Surbhi Suman
Senior Editor: Tanya D'cruz
Technical Editor: Arjun Varma
Copy Editor: Safis Editing
Project Coordinator: Shagun Saini and Deeksha Thakkar
Proofreader: Safis Editing
Indexer: Manju Arasan
Production Designer: Alishon Mendonca
Marketing Coordinator: Nimisha Dua

First published: October 2022

Production reference: 1211022

Published by Packt Publishing Ltd.
Livery Place
35 Livery Street
Birmingham
B3 2PB, UK.

ISBN 978-1-80323-111-2

www.packt.com

A first thank you to my parents: John for intellectually challenging me and Rose for always believing in me (and improving my spelling). To Sam Bansil, as her passion for writing rubbed off on me. And of course, thank you to my beautiful fiancée, Isabella, for her constant stream of love, support, and fresh coffee.

– Zachary Lasiuk

Dedicated to my loving husband, Birju, and my incredible children, Leah and Liam, who would prefer a book about penguins and dragons. Thank you to my parents, Saroj and Praveen, for their endless love and support.

– Pareena Verma

Thank you to my wife, Deborah, my six children (Hannah, Caroline, Philip, Charlotte, Peter, and Maria), their spouses (Stephen, Fernando, and Averi), and my three grandchildren (Max, Thiago, and Agnes) for the encouragement and support.

– Jason Andrews

Contributors

About the authors

Zachary Lasiuk is a solutions designer at Arm. He specializes in being a systems-oriented thinker, with a broad understanding of and experience with the full IoT product life cycle in terms of both hardware and software. He is a designer at heart, crafting products that are easy and enjoyable to use. Graduating summa cum laude from Boston University with a degree in electrical engineering, Zach holds several certifications in fields ranging from UX Design to Design Thinking to Humane Technology. He graduated from the UN Young SDG Innovators Programme and has been an XTC judge for AI Ethics. He enjoys playing jazz saxophone and piano in his spare time and has toured with bands and DJ groups across the world. He lives in Austin, Texas with his fiancée, Isabella, who he loves very much.

Pareena Verma is a principal solutions architect at Arm. She works with Arm partners around the world to design system-level virtual prototyping solutions for early IP evaluation, performance analysis, and software bring-up. She has helped software developers and SoC architects on numerous Arm-based projects involving the usage of modeling, compilers, debuggers, and simulation tools. Prior to working at Arm, Pareena worked at a couple of other Electronic Design Automation start-ups, primarily focused on embedded software development and FPGA design. Pareena holds a Master of Science degree in Electrical and Computer Engineering from the University of Florida. She lives in the Greater Boston area with her husband, Birju, and their two amazing children, Leah and Liam.

Jason Andrews is a solutions director and distinguished engineer at Arm. He helps Arm partners in the areas of IP selection, system architecture, software development, and performance analysis. Jason has written hundreds of articles about Arm technology. As a member of the AWS Community Builders program, he promotes the Arm architecture in cloud and IoT applications. Prior to working at Arm, Jason worked in various Electronic Design Automation companies, including Cadence Design Systems. He lives in the Minneapolis area with his wife, Deborah, where they spend time with their six children and three grandchildren.

About the reviewer

Ronan Synnott is a solutions architect within the Development Solutions group at Arm. During a career spanning more than twenty years, Ronan has held several customer-focused positions around the globe, working side by side with developers from various sectors to ensure they can be at their most effective. He has experienced firsthand how embedded software development has evolved. He is a frequent contributor to Arm's community forums and blogs and at other public events.

Table of Contents

3

Selecting the Right Tools 45

Part 2: Sharpen Your Skills

4

Booting to Main 69

5

Optimizing Performance   97

6

Leveraging Machine Learning   119

7

Enforcing Security   141

8

Streamlining with the Cloud   167

9

Implementing Continuous Integration   193

10

Looking Ahead   217

Index   239

Other Books You May Enjoy   254

Preface

Arm Cortex-M processors are ideal for a wide variety of applications. They are highly visible in microcontrollers and silently work in every other area of electronic design, from small sensors to large servers. In the fourth quarter of 2020, Arm reported a record 4.4 billion chips shipped with Cortex-M processors.

Consequently, the world of software development for embedded and IoT devices is broad. There are hundreds of companies creating thousands of Cortex-M chips, development boards, software libraries, and development tools. While all these components are intended to make your job developing software for Cortex-M devices easier, it is a challenge to understand which components to use on a specific project.

Our goal is to alleviate these challenges and enable you to focus on building better Cortex-M software. We hope our knowledge and experience will help you avoid frustration and spend more time doing what you enjoy.

This book is split into two parts. *Part 1*, *Get Set Up*, focuses on how to select the right components to make a Cortex-M based project successful. We cover which Cortex-M processor makes sense for your application—and hardware options to simplify development. Next is an overview of the large variety of software components available in the Cortex-M ecosystem, with context on when to use them. This part ends with a discussion on embedded software tool selection. After reading *Part 1*, you should be familiar with what exists in the broad Cortex-M ecosystem and be able to translate your project requirements into the right hardware, software, and tools to be successful.

Part 2, *Sharpen Your Skills*, dives into specific topics of Cortex-M software development. We cover both software topics (including system startup, optimization, machine learning, and security) and software development topics (including cloud services and continuous integration testing). Each topic will be explained in theory and in practice, with code examples for you to get experience with along the way. If you are interested in a specific topic, feel free to investigate that chapter sooner; just note later chapters may refer to techniques described earlier in the book.

Who this book is for

This book is intended to teach practicing engineers and students about topics key to being a rounded Cortex-M software developer. It contains some theory and a lot of hands-on examples, as we believe the best way to learn something is to try it yourself. After reading this book, you will have a solid understanding of when and how to use topics such as embedded security and machine learning instead of just knowing the buzzwords.

If you are looking for a deep dive into Cortex-M hardware capabilities or how to optimize Arm assembly instructions for the Cortex-M Instruction Set, technical reference manuals will be best. Similarly, this book is *not* a programming book that explains C or assembly language programming for the purpose of creating a single application or mastering a type of programming such as digital signal processing.

To get the most out of this book, you should have a basic understanding of embedded system hardware and software and general C programming skills.

What this book covers

Chapter 1, *Selecting the Right Hardware*, covers differences between all Cortex-M CPUs, and by extension the SoCs they are packaged in, based on the use cases and performance requirements of a project. Each use case discussion will center around the practical usage of Cortex-M hardware and software support. Hardware features such as FPUs, TCMs, MVE, and TrustZone will be discussed. Software such as ML frameworks, RTOS, and more will be discussed.

Chapter 2, *Selecting the Right Software*, introduces the different software frameworks, ranging from bare-metal boot code to real-time operating systems that enable efficient development on Cortex-M embedded devices. We will map popular use cases to software frameworks available today to get them up and running quickly on your Cortex-M devices.

Chapter 3, *Selecting the Right Tools*, lists the platforms useful for developing products based on Cortex-M hardware. We will list the tools required to work with Cortex-M hardware, with some general commentary to compare them. We will also list the possible development environments people can use, along with the costs and benefits of each.

Chapter 4, *Booting to Main*, covers the basics of booting a bare-metal program on a Cortex-M device. We will walk you through assembly startup code examples and highlight the key bits that need to be programmed to successfully boot the systems. We will introduce scatter and linker files – what they are, how they're structured based on memory layout, and how they're used to link your program to create the final executable. We will also walk you through the different hardware mechanisms that can be leveraged for application input/output, accompanied by relevant examples.

Chapter 5, *Optimizing Performance*, covers different software optimization techniques that can be used to run your code faster and more efficiently on Cortex-M devices. We will discuss the tools and techniques you can leverage to measure the performance gained using these optimization tips.

Chapter 6, *Leveraging Machine Learning*, clarifies how Cortex-M devices process machine learning software. The focus will be on the Cortex-M55, the first Cortex-M designed with ML software in mind. We will introduce the difference between how the Cortex-M55 processes ML instructions as opposed to the other Cortex-M devices through vector processing. Then we will dive into the most popular low-power edge ML applications, speech recognition and image classification, and how to leverage software frameworks to get started quickly.

Chapter 7, Enforcing Security, provides an overview of Arm's TrustZone technology, which provides robust protection and security for IoT devices. We will introduce Trusted Firmware-M and how it can be included and used to build secure firmware for Cortex-M devices. We will also provide software guidelines on topics such as fault detection and exception handling in the context of building secure systems.

Chapter 8, Streamlining with the Cloud, explains the migration of embedded software development to the cloud. We will introduce cloud concepts including source code management, remote code editing, and containers. Examples of various cloud services as well as build-your-own cloud solutions will be provided with software examples to try. Learning cloud development concepts will help you stay up to date with the latest developer trends.

Chapter 9, Implementing Continuous Integration, introduces automated testing, explains why continuous integration is needed, and reviews the challenges it presents. We will explain how to work with various testing frameworks and cloud services and apply them to both physical boards and virtual models. Learning about possible solutions will enable the right level of automation and tools to improve software quality.

Chapter 10, Looking Ahead, provides general tips for how you can go from a good programmer to a great one. Topics for further investigation are presented for you to continue learning. Additional examples of current industry needs are covered, as well as a look forward toward emerging trends and required skills for the future.

To get the most out of this book

We find the best way to learn a topic is to practice it. In the spirit of "learning by doing," there are multiple examples in each chapter in *Part 2*. To make it as easy as possible for you to code along with us, we tried to select freely available software and tools where possible. We spread out examples through Linux and Windows environments, but in many cases, you will be able to use your OS of choice if the tools are supported.

Hardware boards are not free, but the same three platforms are used throughout the book to minimize the amount of hardware you need:

- Raspberry Pi Pico
- NXP LPC55S69-EVK
- Arm Virtual Hardware

The first two boards are self-explanatory. The third option, Arm Virtual Hardware, is not a physical board. It is easily available online and is discussed and used in context at the start of *Chapter 4*.

If you are using the digital version of this book, we advise you to type the code yourself or access the code from the book's GitHub repository (a link is available in the next section). Doing so will help you avoid any potential errors related to the copying and pasting of code.

Download the example code files

You can download the example code files for this book from GitHub at `https://github.com/PacktPublishing/The-Insiders-Guide-to-Arm-Cortex-M-Development`. If there's an update to the code, it will be updated in the GitHub repository.

Download the color images

We also provide a PDF file that has color images of the screenshots and diagrams used in this book. You can download it here: `https://packt.link/vjeD9`.

Conventions used

There are a number of text conventions used throughout this book.

`Code in text`: Indicates code words in text, database table names, folder names, filenames, file extensions, pathnames, dummy URLs, user input, and Twitter handles. Here is an example: "It's clear that the `printf` function is defined in a C header file, `stdio.h`."

A block of code is set as follows:

```
#include <stdio.h>

int main()
{
    printf("Hello Cortex-M world!\n");
}
```

When we wish to draw your attention to a particular part of a code block, the relevant lines or items are set in bold:

```
#include <stdio.h>

int main()
{
    printf("Hello Cortex-M world!\n");
}
```

Any command-line input or output is written as follows:

```
./run.sh -a hello.axf
```

Bold: Indicates a new term, an important word, or words that you see onscreen. For instance, words in menus or dialog boxes appear in **bold**. Here is an example: "Select **System info** from the **Administration** panel."

> Tips or important notes
> Appear like this.

Get in touch

Feedback from our readers is always welcome.

General feedback: If you have questions about any aspect of this book, email us at customercare@ packtpub.com and mention the book title in the subject of your message.

Errata: Although we have taken every care to ensure the accuracy of our content, mistakes do happen. If you have found a mistake in this book, we would be grateful if you would report this to us. Please visit www.packtpub.com/support/errata and fill in the form.

Piracy: If you come across any illegal copies of our works in any form on the internet, we would be grateful if you would provide us with the location address or website name. Please contact us at copyright@packt.com with a link to the material.

If you are interested in becoming an author: If there is a topic that you have expertise in and you are interested in either writing or contributing to a book, please visit authors.packtpub.com.

Share Your Thoughts

Once you've read *The Insider's Guide to Arm Cortex-M Development*, we'd love to hear your thoughts! Scan the QR code below to go straight to the Amazon review page for this book and share your feedback.

https://packt.link/r/1803231114

Your review is important to us and the tech community and will help us make sure we're delivering excellent quality content.

Download a free PDF copy of this book

Thanks for purchasing this book!

Do you like to read on the go but are unable to carry your print books everywhere? Is your eBook purchase not compatible with the device of your choice?

Don't worry, now with every Packt book you get a DRM-free PDF version of that book at no cost.

Read anywhere, any place, on any device. Search, copy, and paste code from your favorite technical books directly into your application.

The perks don't stop there, you can get exclusive access to discounts, newsletters, and great free content in your inbox daily

Follow these simple steps to get the benefits:

1. Scan the QR code or visit the link below

https://packt.link/free-ebook/9781803231112

2. Submit your proof of purchase
3. That's it! We'll send your free PDF and other benefits to your email directly

Part 1:
Get Set Up

These first three chapters provide a broad overview of the options available to build a Cortex-M device, alongside helpful heuristics to narrow down the best options for your project. Confused by the differences in Cortex-M CPUs? Feeling overwhelmed by all the possible software stacks to choose from? Not sure what tools will help you develop software efficiently? We will answer these (and more) questions by reviewing the range of available hardware, software, and tools in the Cortex-M ecosystem.

This part of the book comprises the following chapters:

- *Chapter 1, Selecting the Right Hardware*
- *Chapter 2, Selecting the Right Software*
- *Chapter 3, Selecting the Right Tools*

1
Selecting the Right Hardware

It may be surprising that the first chapter of this book, which is written for Cortex-M software developers, is all about hardware. This is because software, in all its forms, is ultimately run on hardware. It is critical to understand which hardware capabilities exist to properly leverage them in software.

Additionally, you will likely need a development board for debugging your code during development. Some of you reading may even have a level of influence over which hardware is ultimately selected for your device. All in all, no matter what specific situation you are in, understanding what Cortex-M hardware is out there—and what it can do—will help you develop quality software for your current and future projects.

So, in this opening chapter, we will explain how to select Cortex-M hardware and provide an overview of where to find development boards. Note that we will be discussing both individual Cortex-M processors and Cortex-M development boards.

There are different ways to frame which Cortex-M hardware is best suited for your specific project. Examples can be helpful; the first section of this chapter lists common embedded/IoT use cases and presents Cortex-M processors that fit that situation. The side-by-side comparison is also helpful; the second section ranks processors by performance, power, and area metrics. The third section then focuses on development boards, discussing trade-offs.

The chapter ends by selecting two boards that will be used for hands-on examples in future chapters. In a nutshell, the topics we'll discuss in this chapter are presented here:

- Processor selection through use cases
- Processor selection based on performance and power
- Microcontroller development boards

Processor selection through use cases

IoT and **machine learning** (**ML**) applications are not only rapidly evolving and changing the way modern businesses operate but also transforming our everyday experiences. As these applications evolve and become more complex, it is essential to make the right hardware choices that meet application requirements. Ultimately, the processor choice comes down to the right balance of functionality, cost, power, and performance. Defining your use case and workload requirements makes determining this balance a lot simpler.

In this section, we will walk through the requirements of some common consumer embedded use cases and determine the Arm Cortex-M processor choices that are ideally suited. The list of use cases and the resulting processor selections are not exhaustive, mainly highlighting that if workload requirements are well understood, the processor decision-making process becomes much easier.

Medical wearable

Let's start with a smart medical wearable use case. The requirements of this wearable include that it will be a wrist-worn device, with long battery life and special sensors to continuously monitor heart activity. Security is a vital requirement as the wearable stores private medical data. Processing power is equally important, operating within the size and power constraints of a battery-operated wearable.

For this case, the Arm **Cortex-M33** processor provides an excellent combination of security, processing power, and power consumption. Cortex-M33 includes security features for hardware-enforced isolation, known as **TrustZone** for Cortex-M. It reduces the potential for attacks by creating isolation between the critical firmware and the rest of the application. The Cortex-M33 has many optional hardware features including a **digital signal processing** (**DSP**) extension, **memory protection unit** (**MPU**), and a **floating-point unit** (**FPU**) for handling compute-intensive operations. The Arm custom instruction and coprocessor interface in the Cortex-M33 provide the customization and extensibility to address processing power demands while still decreasing power consumption.

Note that these hardware features are optional; once manufactured and sold, these features are either present or not. Make sure to check whether the microcontroller or development board you are buying has these Arm Cortex-M processor features enabled if desired.

Industrial flow sensor

Let's take another use case as an example. Say you're designing an industrial flow sensor that will be used to measure liquids and gases with great accuracy. It needs to be extremely reliable and have a small form factor. The primary requirement is that it will be low-power and work with this accuracy standalone for very long periods of time. A great **central processing unit** (**CPU**) choice for designing such an industrial sensor is the **Arm Cortex-M0+**, which combines low power consumption and processing-power capabilities. It is the most energy-efficient Cortex-M processor with an extremely small silicon area, making it the perfect fit for such constrained embedded applications.

IoT sensor

There are several use cases in the embedded market that require demanding DSP workloads to be executed with maximized efficiency. With the advancements in IoT, there has been an explosion in the number of connected smart devices. There are so many different sensors connected within these devices to collect data for measuring temperature, motion, health, and vision. The sensor data collected is often noisy and requires DSP computation tasks—for example, applying filters to extract the clean data. The **Cortex-M4**, **Cortex-M7**, **Cortex-M33**, and **Cortex-M55** processors come with a DSP hardware extension addressing various performance requirements for different signal-processing workloads. They also have an optional FPU that provides high-performance generic code processing in addition to the DSP capabilities. If your workload requires the highest DSP performance, the Cortex-M7 is a great choice. The Cortex-M7 is widely available in microcontrollers and offers high performance due to its six-stage dual-issue instruction pipeline. It is also highly flexible with optional instruction and data caches and **tightly coupled memory** (TCM), making it easy to select a processor that has been manufactured to meet your specific application needs. Security has become a common requirement for sensors on connected devices to provide protection from physical attacks. If security is an essential requirement for your sensor application in addition to DSP performance, then Cortex-M33 could be a great fit with its TrustZone hardware security features.

With some of the newer sensing and control use cases, we see a common need for not only signal processing but also ML inference on endpoint devices. ML workloads are typically very demanding in terms of computation and memory bandwidth requirements. The significant advancements made in ML via optimization techniques have now made it possible for ML solutions to be deployed on edge devices.

ML

The primary use cases for ML on edge devices today are keyword spotting, speech recognition, object detection, and object recognition. The list of ML use cases is rapidly evolving even as we write this book, with autonomous driving, language translation, and industrial anomaly detection. These use cases can be broadly classified into three categories, as outlined here:

- **Vibration and motion**: Vibration and motion are used to analyze signals, monitor health, and assist with several industrial applications such as predictive maintenance and anomaly detection. For these applications, the installed sensors (generally accelerometers) are used to gather large amounts of data at various vibration levels. Signal processing is used to preprocess the signal data before any decision-making can be done using ML techniques.

- **Voice and sound**: Voice applications are in several markets, and we've become quite familiar with voice assistants through the deployment of smart speakers. Many other voice-enabled solutions are coming to the mass market. The voice-capture process consists of one or several microphones used for voice keyword detection. Keyword spotting and automatic speech recognition are the primary demanding computing operations of these voice-enabled devices. These tasks require significant DSP and ML computation.

- **Vision**: Vision applications are used in several areas for recognizing objects, being able to both sort and spot defects, and detecting movement. There is an increasing number of vision-based ML applications ranging from video doorbells and biometrics for face-unlocking to industrial anomaly detection.

Cortex-M processors ranging from the **Cortex-M0** to the latest **Cortex-M85** can run a broad range of these ML use cases at different performance points. Mapping the different workload performance needs and latency requirements of these use cases to the CPU's feature capabilities greatly simplifies the process of hardware selection. The following diagram illustrates the range of ML use cases run on the Cortex-M family of processors today:

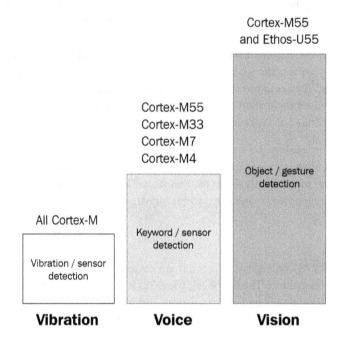

Figure 1.1 – ML on Cortex-M processors

For example, say you're designing a smart speaker that is an always-on voice activation device—the Cortex-M55 is a great choice. The Cortex-M55 is a highly capable **artificial intelligence** (**AI**) processor in the Cortex-M series of processors. It's the first in the Arm Cortex-M family of processors to feature the Helium technology, which provides a significant performance uplift for DSP and ML applications on small, embedded devices. Arm Helium technology is also known as the **M-Profile Vector Extension** (**MVE**), which adds over 150 new scalar and vector instructions and enables efficient computation of 8-bit, 16-bit, and 32-bit fixed-point data types. Signal processing-intensive applications such as audio processing widely use the 16-bit and 32-bit fixed-point formats. ML algorithms widely use 8-bit fixed-point data types for **neural network** (**NN**) computations. The Helium technology makes running ML workloads much faster and more energy-efficient in endpoint devices.

In *Figure 1.1*, there is also mention of **Ethos-U55**. This is not a CPU like the other Cortex-M processors but is instead a micro **neural processing unit** (**NPU**). It was developed to add significant acceleration to ML workloads while being small enough to be implemented in constrained embedded/IoT environments. When combined with the Cortex-M55, the Ethos-U55 provides a 480x uplift in ML performance compared to Cortex-M-based systems without the Ethos-U55! Keep an eye out for microcontrollers and boards that utilize the Ethos-U55, and learn more about it from a high level here: https://www.arm.com/products/silicon-ip-cpu/ethos/ethos-u55.

To summarize this section, one way to select processors is by understanding the use case, clearly defining requirements, ranking them, and identifying project constraints. This is a great place to start the processor selection process.

Next, we will look at using performance and power as metrics to analyze processor selection choices.

Processor selection based on performance and power

Another way to choose and understand Cortex-M processors is by ranking, based on how well they match performance and power requirements. Without structure, this can be a daunting task, with a wide range of possibilities (from the number of interrupts to the overall price, and everything in between). In this section, we will define six categories to evaluate and go over a few examples of how to use this in practice to select the right processor for your project. Again, this is also a helpful framework for understanding what Cortex-M processors' capabilities are.

We will select the right processor using an approach we will call *requirement heuristics*. This means translating your key project requirements into predefined areas and following simple steps to get to the right Cortex-M processor. The six areas are listed here:

- Power
- DSP performance
- ML performance
- Security
- Safety
- Cost

In each area, we rank the processors that best meet the project requirement. You can then select the areas that matter most to your project and find the processor that meets these needs. Let's discuss each area before showing some examples.

Power

Minimizing power consumption is crucial in highly constrained power environments. A common use case is in distributed sensors that require long periods of continuous operation without being serviced.

When looking at power metrics, there are a few things that will help in understanding technical jargon. There are often two types of power measurements: static (or leakage) power consumption and dynamic power consumption. Static power consumption measures the amount of power used by the processor when not actively processing anything, such as being in a "sleep" mode but with the power still on. Dynamic power consumption measures the power consumed when the processor is actively working on a task. Often, dynamic power is measured using an industry-standard software workload called *Dhrystone*, to enable consistent comparisons. Power is measured in **microwatts (uW)/megahertz (MHz)**. It is defined as power per MHz to enable consistent comparisons between processors running at different frequencies.

The following table shows the dynamic power of the different Cortex-M processors on the same node size:

Arm Cortex-M0+

Dynamic Power	3.8µW/MHz

Arm Cortex-M23

Dynamic Power	3.86µW/MHz

Arm Cortex-M4

Dynamic Power	12.26µW/MHz

Arm Cortex-M33

Dynamic Power	12.0µW/MHz

Table 1.1 – Dynamic power across different Cortex-M processors on 40 LP node size

Another factor that affects the power consumption of the core is the technology node size used to manufacture the silicon. This can be referred to as "technology," "process node," or just "node," depending on the context. The node size refers generally to the physical size of the transistors; as the node size decreases, more transistors can be packed onto the same-size silicon wafer. Smaller node sizes also generally lead to reduced power consumption, both static and dynamic. Understanding the node size is helpful for accurately comparing different chips or boards.

Using a consistent node size of 40 nm and publicly available benchmarking data, we can rank the Cortex-M processors on the low-power axis, like so:

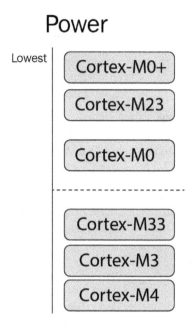

Figure 1.2 – Ranking power consumption for Cortex-M

Note that the processors in the preceding screenshot are ordered from best, starting from the top. In this case, the processors requiring lower power for operation are ranked visually higher. We are also not displaying all the Cortex-M processors in this (and subsequent) screenshots—we're only displaying processors that perform well in the category. The spacing between processors is not intended to communicate precise quantitative differences, only a general ranking based on power consumption. The dotted line is also intended as a qualitative distinction, indicating a notable separation between the processor capabilities on this axis.

The Cortex-M0+ comes in first, as one of the lowest-power 32-bit processors on the market. It can get as low as 4 uW/MHz in dynamic power consumption when manufactured at 40 nm. This processor is being used at the forefront of low-power technologies. It can even be used in applications without a battery, relying on energy harvesting from the environment to power the device. Now that is low power! We talk about energy harvesting and ultra-low-power applications in *Chapter 10, Looking Ahead*.

The Cortex-M23 is essentially tied with the Cortex-M0+ in terms of minimizing power. It can achieve similar power figures as the Cortex-M0+ when configured minimally. The security features increase power consumption when included. Overall, given how new the Cortex-M23 is and that it is being used more often at lower node sizes such as 28 nm and below, the Cortex-M23 is equally viable for minimizing power consumption.

The Cortex-M0 also minimizes power draw and is only slightly behind the Cortex-M0+ and Cortex-M23. The Cortex-M0+ is typically a better option than the Cortex-M0, being so closely related.

The Cortex-M33, Cortex-M3, and Cortex-M4 all have about triple the power draw as the Cortex-M0+. If the lower-power-consumption processors do not have enough processing power or features for your use case, these processors are likely a good fit.

DSP performance

DSP is needed when taking real-world signals and digitizing them to perform some computations. This is exceedingly common in movement, image, or audio processing applications when data is coming in real time. Devices with sensors and motors to detect and act on real-time data rely heavily on DSP.

The computational nature of these types of DSP applications is really centered on what we call *scalar* processing. You may be familiar with the word *scalar* from math or physics classes. A scalar is a quantity that has only one characteristic. For example, measuring gas pressure 10 times a second for 1 second will produce 10 data points. Each point has one characteristic: the magnitude of the gas pressure at that instant. These types of DSP applications, which include audio processing as well, lend themselves well to scalar processing.

To measure how good Cortex-M processors are at scalar processing, there are two common benchmarks: CoreMark and Dhrystone. Using these imperfect but generally helpful benchmarks, you can compare how well different processors run scalar workloads such as the DSP use cases discussed previously. You can download and view the Dhrystone (**Dhrystone Million Instructions per Second (DMIPS)/MHz**) and CoreMark scores for all Cortex-M series processors here: https://developer.arm.com/documentation/102787/.

Using these publicly available CoreMark benchmark scores compiled with the Arm Compiler for Embedded, we can rank the Cortex-M processors in terms of DSP performance, as follows:

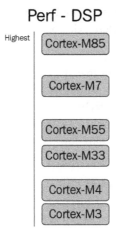

Figure 1.3 – Ranking DSP performance for Cortex-M

The benchmark numbers quoted next are valid at the time of this book's publication. Due to subtle changes in firmware, benchmarks, and compilers, the numbers may change slightly over time. These changes will be small, and the rankings listed are still directionally accurate.

As the newest Arm Cortex-M processor, the Cortex-M85 provides the highest scalar and signal-processing performance to date in the Cortex-M family. It boasts a CoreMark score of 6.28 CoreMark/MHz, is suitable for the most demanding DSP applications, and also includes TrustZone security features.

The Cortex-M7, while being superseded by Cortex-M85, is still a good choice for less demanding DSP applications or where functional safety is critical. The Cortex-M7 has a CoreMark score of 5.29.

The Cortex-M55 and Cortex-M33 are similar in scalar performance, with a CoreMark score of 4.4 and 4.1 respectively.

The Cortex-M4 and Cortex-M3 are the next steps down in performance, with CoreMark scores of 3.54 and 3.44 respectively. The Cortex-M4 is better with DSP use cases due to its optional FPU (which the Cortex-M3 does not have). The Cortex-M4 is commonly used in sensor fusion, motor control, and wearables. The Cortex-M3 is used for more balanced applications with lower area and power requirements.

Applications involving video processing are more demanding than traditional DSP software and benefit from simultaneous processing, called vector processing. Vector processing accelerates the most popular workload today—ML.

ML performance

Because of its increased popularity and potential in edge devices, we will devote an entire chapter to ML in *Chapter 6, Leveraging Machine Learning*. In this section, we will give an overview of how ML workloads are executed in hardware to identify the right Cortex-M processor for the job.

ML, at a computational level, is matrix math. NNs are represented by layers of neurons, with each neuron in one layer being connected to each neuron in the next layer. When an input is given (such as a picture of a cat to an image recognition network), it gets separated into distinct features and sent through each layer, one at a time. In practice, this means at each layer, there is x number of inputs going into n number of nodes. This leads to $x*n$ computations at each layer, of which there could be dozens, with potentially hundreds of nodes in each layer.

In scalar computing, this could result in tens of thousands of calculations performed one after the other. In vector computing, you instead store each node's value in a row (or lane) and make $x*n$ calculations all at once. This is the benefit of vector processing, which has existed in larger Arm cores for years via NEON technology. The Helium extension brings this technology to Cortex-M processors without significantly increasing area and power.

Using matrix multiplication performance as a benchmark, we can rank the Cortex-M processors in terms of ML performance, like so:

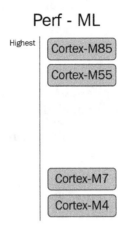

Figure 1.4 – Ranking ML performance for Cortex-M

The Cortex-M85 processor boasts the most recent implementation of the Helium vector processing technology. It brings more ML functionality to edge devices and enhances applications such as robotics, drones, and smart home control.

The Cortex-M55 processor was the first Cortex-M processor with Helium vector processing technology. It brings anomaly and object detection use cases to the edge when implemented standalone. When paired with an NPU (such as the Ethos-U55 discussed earlier in this chapter), gesture detection and speech recognition use cases can be unlocked while still controlling power consumption and cost. Even by itself, the Cortex-M55 has about an **order of magnitude (OOM)** better ML performance than the next closest, the Cortex-M7.

The Cortex-M7 processor is a superscalar processor, meaning it enables the parallelization of scalar workloads. This effectively allows it to run DSP applications faster, but the more computationally intensive ML use cases are more of a challenge. This processor is suitable for basic ML use cases such as vibration and keyword detection.

The Cortex-M4 processor is often stretched to its computational limits when applied to ML use cases. In most cases, it should only be considered if the ML use case is around vibration/keyword detection or sensor fusion, and there is a strict power or cost constraint.

Security

As the importance and ubiquity of IoT devices have increased in people's lives, security has become a strong requirement. The basics of security such as cryptographic password storage are no longer acceptable. As the value and volume of what is stored on edge devices increases, malicious actors get proportionally more incentivized for hacks.

We will also devote *Chapter 7, Enforcing Security*, to the topic of security and dive into more specifics there. This section will give you an overview of the key considerations to remember when selecting a processor with security in mind. To successfully secure your software and project, the underlying hardware needs to enable some essential features such as software isolation, memory protection, secure boot, and more. Arm has implemented a security extension to the newer Cortex-M processors called TrustZone that enhances these security basics, adds more functionality in hardware, and makes security implementations easier. TrustZone enables you to physically isolate sections of memory or peripherals at the hardware level, making hacks more difficult and more contained if they do occur. The full details, benefits, and a quick-start guide for this extension will be provided in a later chapter.

Note that this is an optional extension, so make sure to verify it is enabled for the processor in any development board you are considering.

Using TrustZone and additional security features as a guide, we can rank the Cortex-M processors in terms of security features, as follows:

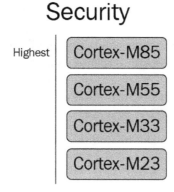

Figure 1.5 – Ranking security for Cortex-M processors

In practice, these processors all contain the TrustZone security extension and are all excellent options for developing a secure project. They are all based on the Armv8-M instruction architecture, with the other Cortex-M processors being based on Armv7-M or Armv6-M. They are ordered in terms of most recently released, but other requirements such as low power, ML, or DSP performance should decide which of these processors to select. Note that Arm also has a **Platform Security Architecture (PSA)** certification that validates the security implementations at the development-board level.

The PSA and TrustZone implementations in software are all discussed more in *Chapter 7, Enforcing Security*. Resources to learn more about the different Arm instruction sets (outside the scope of this book) are listed under the *Further reading* section at the end of this chapter.

> **Important note**
>
> The **Cortex-M35P** processor is a specialized processor that is intended for the highest level of security. It features built-in tamper resistance and physical protection from invasive and non-invasive attack vectors. Basically, it is ideal for devices that protect valuable resources but are accessible by the public, such as a smart door lock. The core is similar to the Cortex-M33, adding that physical layer of protection. If your product needs physical, tamper-proof security as the primary requirement, this is definitely a Cortex-M processor to consider.

Safety

In the embedded space, safety requirements typically come when in a regulated, safety-critical environment. These safety requirements can be categorized as either diagnostic requirements or systematic requirements. Diagnostic requirements relate to the management of random faults on the device and are addressed by the addition of hardware features for **fault detection** (**FD**) and control. Systematic requirements relate to demonstrating the avoidance of systematic failures and are addressed typically through the design process and verification.

Products sold to high-safety environments must prove a level of risk reduction as defined by international standards. The **International Electrotechnical Commission** (**IEC**) *61508* standard defines general **Safety Integrity Levels** (**SILs**), with SIL 4 being the most strict and SIL 1 being the least strict. The automotive industry has a dedicated level system, the **Automotive Safety Integration Level** (**ASIL**), with ASIL D being the strictest and ASIL A being the least strict.

Some common safety features include the following:

- Exception handling, which prevents software crashes in the case of system faults.
- MPUs that ensure data integrity from invalid behavior.
- **Software test libraries** (**STLs**) test for faults at startup and runtime. Note that this is not a feature of the processor, but instead, a suite of software tests provided to run on a specific processor.
- **Dual-Redundant Core Lockstep** (**DCLS**), where two processors redundantly run the same code to uncover and correct system errors.
- **Error Correction Code** (**ECC**), which automatically detects and corrects memory errors.
- **Memory Built-In Self-Test** (**MBIST**) interfaces that enable memory integrity validation while the processor is running.

We can rank the Cortex-M processors in terms of safety features, showing the cutoff for which processors are capable of reaching certain safety levels, as follows:

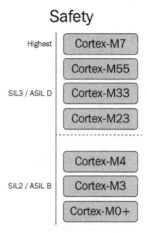

Figure 1.6 – Ranking safety features of Cortex-M processors

The Cortex-M7 is alone in the Cortex-M family in offering ECC, MBIST, and DCLS features alongside the more common MPU and exception handling. The Cortex-M55, Cortex-M33, and Cortex-M23 contain almost all of those features, but are still capable of meeting the strict SIL 3 and ASIL D safety levels.

The Cortex-M4, Cortex-M3, and Cortex-M0+ all offer enough safety features to achieve the least strict SIL 2 and ASIL B safety levels with STLs, MPUs, and exception handling.

The Cortex-M35P processor is highly effective for safety applications as well as security applications. It contains most of the already listed safety features, adding in heightened observability to ensure expected behavior and more.

Now that we have looked at some key features that can drive your processor selection, let us look at how cost can impact this decision-making process.

Cost

Minimizing cost is a common requirement in deeply embedded and IoT spaces. When looking at microcontrollers or boards to purchase, the cost should be obvious and does not require much explanation.

We can, however, provide some context for what contributes to the cost of a microcontroller, with the largest factor here being the silicon area. As the area of a microcontroller increases, it requires more materials to make and thus intuitively raises costs. Production volume will also impact the cost. The higher the production volume, the lower the cost will be. Typically, the smaller the Cortex-M processor is, the less expensive it will be to manufacture, and thus the lower the price to purchase. We will review the Cortex-M processors with the lowest area so that you have a starting point to look for boards with these processors to have the best chance of minimizing your overall cost.

> **Important note**
>
> These Cortex-M processors are highly configurable, and the implementation of more features will increase the area and likely increase cost. In looking for a microcontroller or board while minimizing cost, make sure to select a product with only the minimum set of features you need to hold down the cost as much as possible.

Here are the Cortex-M processors ranked in terms of lowest possible area:

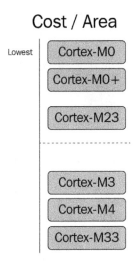

Figure 1.7 – Ranking area of Cortex-M processors

The Cortex-M0 and Cortex-M0+ are tied at the lowest area and are commonly found at the lowest price points. These are excellent choices for low-cost applications if they have enough functionality.

The Cortex-M23 is just behind the Cortex-M0 and Cortex-M0+ in terms of area. The Cortex-M23 has the benefit of enhanced security features, making it a great choice for low-cost connected use cases.

The Cortex-M3, Cortex-M4, and Cortex-M33 all follow, being notably larger than the previous Cortex-M processors. These are the next best options to look at when you need more functionality while keeping costs low.

With this background about selecting Cortex-M processors, let's look at how to find development boards that include Cortex-M processors.

Microcontroller development boards

Now that we have introduced Cortex-M use cases and guidance on making processor selections based on use cases and other factors such as performance and power, let's find some development boards and get started learning.

Numerous development boards are available with Cortex-M microcontrollers. While it's impossible to provide a simple flow chart for how to select a Cortex-M microcontroller, we can provide an overview of the options and trade-offs as a starting point for narrowing the selection.

At the end of this chapter, we will pick two development boards and use those to demonstrate the concepts in the remainder of the book.

Suppliers

There are multiple suppliers for Cortex-M microcontrollers, and each supplier typically offers a wide range of options to optimize for specific markets and use cases. It's common for a microcontroller supplier to offer hundreds of Cortex-M devices. Each supplier also offers a range of development boards to try out the devices and get started on projects quickly. Here is our recommended list of suppliers to get familiar with:

- NXP
- Infineon
- STMicroelectronics
- Nordic Semiconductor
- Raspberry Pi
- Nuvoton
- Renesas
- Silicon Labs
- Analog Devices
- Dialog Semiconductor

Microcontrollers and development boards are normally sold by distributors around the world. Some of the common distributors include the following:

- Arrow
- DigiKey
- Arduino
- CanaKit
- Mouser
- Adafruit

Popular boards may also be available online at places such as Amazon and Newegg.

Now that we know the microcontroller suppliers and where to find development boards, let's review the primary factors to consider when selecting a microcontroller and a board.

Selecting development boards

In the previous sections, we reviewed several criteria for selecting a processor, as follows:

- Use cases
- Power
- DSP performance
- ML performance
- Security
- Safety
- Cost

Selecting microcontrollers and development boards introduces a few new factors that impact performance and functionality. These include memory characteristics, power supplies, and peripherals.

Performance

Performance is driven by the compute requirements of software applications. Selecting hardware is always easier when the software is well understood, but it's not always possible. When software is not available, similar applications can be used to estimate performance. There are a variety of industry benchmarks that may provide performance guidance.

As we have seen in previous sections, microcontroller performance is primarily determined by the Cortex-M processor microarchitecture. The next most important factor is memory.

Important CPU factors include the frequency, instruction pipeline, cache, TCM, floating-point hardware, and any special instructions for DSP or vector processing of larger units of data.

Different types of memory are also available in microcontrollers. Most microcontrollers use a mix of **static random-access memory** (**SRAM**) and flash memory. Flash memory can be reprogrammed numerous times and holds data when the power is removed. Flash is ideal for storing software instructions.

SRAM provides faster access compared to flash, but the contents are lost when power is removed.

Dynamic RAM (**DRAM**) may also be used in embedded systems when larger amounts of memory are required, and SRAM would be too expensive.

A common software strategy is to move important code and data from flash to SRAM after analyzing to improve performance.

Cortex-M microcontrollers are easy to use because they have a standard memory map that places **read-only memory** (**ROM**) or flash memory and SRAM at standard locations in the memory map. This makes it easy to set up regardless of the Cortex-M processor being used.

Looking at the memory access times is key to analyzing performance. Microcontrollers may contain caching systems to feed data from flash memory into a CPU much faster than making a single read access to flash and incurring waits. Instruction and data caches also improve access time to memory.

Memory size is also a key parameter for selecting a device. Most Cortex-M systems are memory-constrained. Running out of memory can be a challenge, so keep an eye on the size of each memory type.

Power

Power is driven by the operating environment of a product. Requirements to run on a battery, use a specific type of power supply, or maximum current and voltage values to work in each environment should be considered. Certainly, battery versus always plugged in makes a difference, but power consumption has become more complex than just leaving products plugged in and forgetting about power. Products and consumers are starting to focus on using less energy, even in IoT products that are always plugged in, because saving energy across a trillion IoT devices makes a significant difference. Consumers are starting to ask whether every IoT device needs to continuously monitor and report data even while they are sleeping.

Arm processors have a long history of being efficient and using a reasonable amount of power for the performance they provide. Cortex-M processors emphasize simplicity and low-power sleep modes when processing is not needed.

Development boards will provide technical specifications with power details. Using the board specifications in combination with what we have learned about individual Cortex-M processors will help with making good decisions about power.

Peripherals

One key distinction between the thousands of Cortex-M microcontrollers is the peripherals of each device. In fact, this is a common starting point when narrowing down options to make sure desired functionality is supported. The cost-sensitive nature of microcontrollers leads each vendor to develop families of products, selection guides, and tables of products that mix and match various factors such as processor features, peripherals, memory sizes, and temperature ranges. The tables of options can be overwhelming.

One way to understand peripherals is to break them down into a few categories. Today's IoT devices perform data collection, processing, and data communication. Historically, embedded systems focused on **input/output** (**I/O**) pins to connect many types of hardware for data and control.

You can segment peripherals into categories through the following:

- Input from the environment (collecting data from sensors)
- Communication interfaces (storing data)
- Digital and analog I/O signals to interface with additional hardware

A useful exercise to get an understanding of peripherals is to scan through a variety of microcontroller technical summary descriptions and see what is highlighted. These are the key factors the device vendor decided to highlight. This includes a combination of unique features and common features that many devices may have.

Here are some key highlights we found when looking across a variety of devices.

Cortex-M0 and Cortex-M0+ development boards

Some of the common features advertised for Cortex-M0 and Cortex-M0+ boards include **general-purpose I/O (GPIO)** pins and analog input pins. These allow projects to connect other digital and analog hardware to the board. A **Universal Serial Bus (USB)** is also common on most development boards for easy connection to software development tools. Some basic sensors may be included for things such as temperature or other climate sensors. Because of the focus on cost and simplicity, another key feature is the types of expansion connectors that are available. This allows users to expand in a generic way to add more interfaces for both data collection and communication to store data. Standardization of connectors brings better compatibility and a wider selection of add-on hardware.

Cortex-M3, Cortex-M4, and Cortex-M7 development boards

Development boards with these Cortex-M processors highlight IoT. Communication features to store data remotely are highlighted, so the boards can immediately be used for IoT applications. Common technologies are **Bluetooth Low Energy (BLE)**, **near-field communication (NFC)**, **radio frequency (RF)**, and Wi-Fi. This class of boards also highlights more advanced sensors such as accelerometers and gyroscopes, as well as **pulse width modulation (PWM)** to control lights and motors. **Analog-to-digital converters (ADCs)** and **digital-to-analog converters (DACs)** are also common to interface with a variety of digital and analog hardware.

Many applications will be battery-powered, so battery types and means to power the boards are important.

Cortex-M23 and Cortex-M33 development boards

Cortex-M23 and Cortex-M33 development boards emphasize TrustZone features and support for **Trusted Firmware-M (TF-M)** to provide a trusted environment for secure storage and cryptography.

Interfaces for these development boards will be similar to the previous group but can be more advanced as the processors are newer and security features enable new applications. Communication interfaces include BLE, Wi-Fi, and **Long-Term Evolution (LTE)** modems. Specialized antennas may be available for advanced communication protocols such as a **NarrowBand-IoT (NB-IoT)** antenna. More complex features such as a stereo audio codec are common. Additional storage is possible using a microSD card slot.

Some development boards in all categories include a connector to measure the power of a running system. This helps to make it easy to measure actual power during system development.

Other factors

In addition to analyzing the Cortex-M processors, memory types, and peripherals, there are several secondary factors to consider when selecting microcontrollers and development boards.

Software

The most important factor is software, which is why we are writing this book. There are numerous software considerations, and these will be covered specifically in the next chapter. Topics are **real-time operating system (RTOS)** support, connectivity standards, **software development kit (SDK)** options and programming languages, and software libraries.

Development boards can also be certified by **cloud service providers (CSPs)** or serve as primary reference boards, for example, software to make it easier to get going right away with a particular RTOS, connectivity standard, or IoT platform for data collection and visualization.

When possible, it's also good to use benchmark or representative software that is similar to the application to guide performance and code size in selecting the right development board.

Support community

An important secondary factor when selecting development boards is the ability to find answers to questions. Because there are so many microcontrollers and development boards, some of them may not have a large or active community of users reporting problems and helping each other find answers. This often makes a significant difference in how fast a new development board can be adopted. If you believe having an active community around a board is helpful, try to search for a representative problem and see whether there are answers available. Some development boards just die out, while others seem to have active developer forums. Raspberry Pi has a large following, and there is so much content it feels like somebody has already tried almost every possible thing to be done with the boards. This makes Raspberry Pi boards easy to learn.

Size

Most microcontrollers themselves are physically very small, but the size of the development board could be very small or somewhat larger. If you want to use a development board to prototype an idea for a new product, it is good to know whether a prototype can be created that is close enough to demonstrate the product concept. Even if a custom board will be created later, having a prototype of useful size may accelerate product development.

Temperature range

Microcontroller products may have temperature ranges assigned to them. Historically, devices that operate over a range of 0°C to 70°C or 85°C are considered consumer-grade. A temperature range of -40°C to 85°C is considered industrial-grade. There may also be special characteristics for automotive microcontrollers. It's worth checking into these extended temperature ranges if needed, but extending the temperature range comes at a higher price. Industrial IoT is popular and a common place to find extended temperature microcontrollers.

Availability

As we write in 2022, the global electronics industry has been suffering from a *chip shortage*. This has made it harder to find development boards at the various distributors. Having a good handle on availability helps us to know whether more development boards are needed for additional engineers or to build a test lab. Make sure to spend a little time understanding the availability of microcontrollers and development boards before getting too deep into a project. We hope that in the months and years to come, availability will be much less of a factor.

Development board selections

For the upcoming chapters, we selected two boards to demonstrate the concepts and provide detailed steps for you to try things yourself. Trying out the concepts presented in the book is the best way to learn and have the most fun. We selected two boards to cover small Cortex-M devices, security features of Cortex-M, devices with peripherals, and connectivity, as follows:

- **Board 1**: Raspberry Pi Pico using the Cortex-M0+
- **Board 2**: NXP LPC55S69-EVK using the Cortex-M33

We will also use a virtual board to demonstrate the concepts of the Cortex-M55.

We tried to avoid boards that are very new or may be difficult to obtain. We also tried to select boards with reasonable prices that can demonstrate newer concepts such as security and network connectivity.

Summary

Thankfully, microcontroller development boards are inexpensive, and it's easy to try a variety of them. If you end up with the wrong CPU performance, not enough memory, or need another peripheral, it's not difficult to do some more research and try another board.

In this chapter, we provided a survey of important factors to consider when selecting Cortex-M processors, microcontrollers, and development boards. We discussed several criteria for selecting a processor, ranging from use cases to power, performance, security, safety, and cost.

Finally, we proposed two development boards to use in future chapters for hands-on activities. Let us move into *Chapter 2*, *Selecting the Right Software*, where we will learn about the broad range of software available for Cortex-M microcontroller projects.

Further reading

For more information, refer to the following resources:

- Cortex-M processor comparison table: `https://developer.arm.com/documentation/102787/`

- Cortex-M55 Helium explanation: `https://developer.arm.com/Architectures/Helium`

- Ethos-U55 resources to understand more about NPUs: `https://developer.arm.com/Processors/Ethos-U55`

- Armv8-M architecture reference manual: `https://developer.arm.com/documentation/ddi0553/`

- Armv7-M architecture reference manual: `https://developer.arm.com/documentation/ddi0403/`

- Armv6-M architecture reference manual: `https://developer.arm.com/documentation/ddi0419/`

2
Selecting the Right Software

Arm Cortex-M processors can execute a diverse range of software. The term "software" is so broad in the embedded and **Internet of Things** (**IoT**) industries due to the widely varying capabilities of Cortex-M processors. Firmware, middleware, libraries, and components all refer to different parts of the software stack that make up an embedded device. Further complicating matters is a lack of naming consistency, with people often referring to the same software part by different terminology. We try to use the most common software vernacular in this book.

This chapter aims to provide an overview of the different types of software commonly used in microcontroller applications, alongside information about when they should be used. Similar to the previous chapter about understanding which hardware is available in the ecosystem, this chapter presents the same context for software. It is not intended to deeply explain how each software component works with examples; instead, the chapters in *Part 2, Sharpen Your Skills*, will dive deeper into select software components in key areas.

We start by comparing and contrasting bare-metal software and **real-time operating systems** (**RTOSs**) as the foundational software for any microcontroller application. Next, we review a variety of time-saving, reusable middleware and software libraries. This software provides access to hardware in a consistent way and is optimized for performance. Software reuse saves time and provides the best performance when creating applications by providing interfaces to control hardware. We also look at **machine learning** (**ML**) frameworks for Cortex-M, a rapidly growing application space.

The chapter then explains how to utilize TrustZone for Cortex-M on processors that support this feature, enabling secure software execution. A number of use cases (the same from *Chapter 1, Selecting the Right Hardware*) are also presented that map the available software components to an application so that you can see how they fit together to create a complete microcontroller application.

The chapter ends by surveying **software development kits** (**SDKs**) from microcontroller vendors to get up and running with many of these software components quickly and easily.

The following questions will be addressed in this chapter, to provide you with a more complete understanding of Cortex-M software offerings and possibilities:

- What is bare-metal software?

- What is an RTOS?

- What are common middleware and libraries for Cortex-M devices?

- What options exist to implement safe and secure code?

- What does a reasonable software stack look like for common use cases?

- Which SDKs can be used for Cortex-M devices?

The first choice for many microcontroller applications is whether to use an RTOS or use bare-metal software as the foundation of the application. In the following section, we will highlight the distinguishing characteristics of bare-metal software and RTOSs. We will also provide an overview of when it is appropriate to use one over the other.

Overview of bare-metal software

The term *bare-metal software* refers to code with no **operating system** (**OS**) or **application programming interface** (**API**). It is written directly onto hardware, thus the name of writing software on the "bare-metal" hardware.

If implementing your application is a straightforward task, then bare-metal software makes sense and is a good place to start developing any application. Bare-metal tasks are either polled or triggered by interrupts. More complex scheduling functionality is enabled through an OS. While OSs can be overly complex for simple applications, they introduce a necessary abstraction and management layer needed for most applications. Increasingly, Cortex-M devices perform many tasks (reading sensors, processing sensor data, transmitting data, and more) that can be managed easier by an OS.

Note that you can add functionality to your bare-metal system through middleware stacks—which will be covered later in this chapter—but implementing middleware is typically easier with an OS.

In general, if your application is simple and there are not a lot of tasks, bare-metal software is a good option.

Overview of RTOSs

RTOSs were created as embedded devices matured to simplify software development—like OS-based software—while retaining the reliability and deterministic behavior of bare-metal software. The more sophisticated microcontroller-based products with their variety of peripherals benefit from the multitasking, deterministic behavior and services of an RTOS. RTOSs come with a scheduler that makes it easier to manage a large variety of tasks. The main difference between an OS and an RTOS is that an RTOS responds to external events in deterministic and minimal time—hence the name *real time*. Middleware stacks are often available with RTOSs and can be easily integrated.

There is a wide variety of RTOSs available in the ecosystem for Arm Cortex-M-based devices. Some of these are available for free and are open sourced, while others are available at cost for commercial use and often packaged with an **integrated development environment** (IDE) from the vendor. We have listed some of the most popular ones here, along with a link to where you can get them, in alphabetical order:

- AliOS Things: `https://github.com/alibaba/AliOS-Things`
- Amazon FreeRTOS: `https://github.com/aws/amazon-freertoshttps://github.com/aws/amazon-freertos`
- Azure RTOS: `https://github.com/azure-rtos`
- Keil RTX5: `https://github.com/ARM-software/CMSIS_5/tree/develop/CMSIS/RTOS2/RTX`
- Mbed OS: `https://github.com/ARMmbed/mbed-os`
- RT-Thread: `https://github.com/RT-Thread/rt-thread`
- Zephyr: `https://github.com/zephyrproject-rtos/zephyr`

Step 1 when selecting a software stack for Cortex-M-based devices is to decide between a base of bare metal or an RTOS. Step 2 is to understand your functional requirements, such as connectivity or heavy **digital signal processing** (DSP), and select software libraries that simplify their implementation. The next section will cover the middleware and libraries to keep in mind in common situations.

Exploring middleware and libraries

To keep up with modern embedded design requirements, microcontrollers offer a wide range of peripherals such as a **Universal Serial Bus** (USB), Wi-Fi, **Bluetooth Low Energy** (BLE), and so on.

Developing software from scratch that utilizes these peripherals efficiently presents developers with real challenges. Easy-to-use middleware helps bridge this gap to properly leverage modern communication and interface peripherals.

Most of the RTOSs we listed in the earlier section offer a rich set of modules and middleware to help software developers easily integrate support for these peripherals in their Cortex-M devices. The following diagram shows some examples of common middleware components:

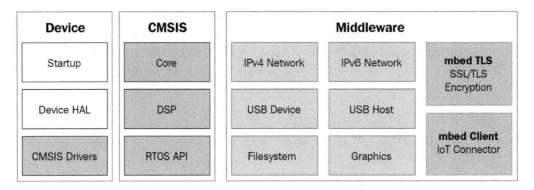

Figure 2.1 – Middleware landscape

As there are so many different types of peripherals that do the same thing (Wi-Fi chips, BLE peripherals, and so on), it can feel like an overwhelming task to find the specific middleware or libraries that accomplish what you need. However, there is a common standard used by the Cortex-M ecosystem that substantially simplifies this process.

CMSIS

The **Cortex Microcontroller Software Interface Standard (CMSIS)** is a collection of API definitions, libraries, utilities, and methods that simplify your software development on Cortex-M devices. It is open sourced and available here for download: `https://github.com/ARM-software/CMSIS_5`.

There are several components in the CMSIS library and several resources available online that provide details on each of these components. This blog on the Arm community site provides a great overview of CMSIS: `https://community.arm.com/arm-community-blogs/b/tools-software-ides-blog/posts/which-cmsis-components-should-i-care-about`. While we are not going into the details of each software component in CMSIS, we will highlight some components that are widely used and their purposes.

System startup, processor core access, and peripheral definitions are essential for every embedded application. With bare-metal programming, you need to provide startup files to perform hardware initialization, C library initialization, and set up memory stack and heap spaces.

CMSIS-CORE, a component of CMSIS, provides the startup code, system configuration files, and device header files for all Cortex-M processors.

Startup code performs several things before entering the main user application, such as the following:

- Sets up reset and exception vector handlers
- Sets up and enables **memory protection unit (MPU)** (when available in the processor)
- Prioritizes and enables interrupts

System configuration performs the processor clock setup, and device header files provide access to the processor core and all its peripherals including the **nested vector interrupt controller** (**NVIC**) and **System Tick Timer** or **System Time Tick** (**SysTick**), which configures periodic timer interrupts.

Now that we have an understanding of what CMSIS is, let us take a look at some of the specialized components in CMSIS that can be leveraged for developing DSP algorithms and ML applications.

DSP libraries

Applications that rely on DSP are generally time-critical and usually involve complex mathematical operations. Because of this, developing real-time DSP systems is far from trivial.

CMSIS-DSP, another component of CMSIS, is a software library of more than 60 algorithms for common signal-processing functions that can be downloaded here: `https://github.com/ ARM-software/CMSIS_5/tree/develop/CMSIS/DSP`.

This collection of common signal-processing functions is categorized into the following functions:

- **Basic math functions**: Mainly consists of optimized functions that streamline vector processing. Vector addition, subtraction, multiplication, and dot product are among the simple math operations. Vector negate, scale, shift, and absolute value allow other types of common data manipulations.

- **Fast math functions**: Offers quick approximations of common math functions that are more involved such as sine, cosine, square root, and division.

- **Complex math functions**: Provides easy manipulation of numbers with imaginary components with functions such as complex dot product, magnitude, and multiplication.

- **Filtering functions**: These are explained in more detail here:

  - **Image processing**—Basic filters such as convolution and correlation are included that enable simple signal manipulations such as sharpening or blurring.

  - **Audio processing**—Several biquadratic functions are included that are commonly used to offer audio filters such as low-pass, high-pass, band-pass, low-shelf, high-shelf, peaking/ bell, and all-pass.

  - **Video processing**—Several **finite impulse response** (**FIR**) filters are included that are fundamental operations for video (or image) processing. Enables contrast improvement, denoising, sharpening, feature enhancement, and more.

- **Matrix functions**: Similar to the basic math functions but specifically for matrix operations. Matrix addition, subtraction, transposing, inverse, initialization, and more make working with matrices simple.

- **Motor control functions**: Functions that enable control of physical motors such as for robotics. The **proportional integral derivative** (**PID**) motor control function is the most commonly used in this category.

- **Transform functions**: Offers optimized real and complex **fast Fourier transform** (**FFT**) functions, which are exceptionally common in audio processing applications.

- **Statistical functions**: Provides basic functions common to statistical analysis, such as finding the maximum, minimum, mean, **root mean square** (**RMS**), standard deviation, and variance from a number set.

- **Support functions**: Help in manipulating stored values into the required formats to simplify computation. They enable converting 8-, 16-, and 32-bit numbers (integer or vector) into smaller or larger sizes. Vector fill, copy, and sorting functions enable streamlined vector usage as well.

- **Interpolation functions**: Can be used in various DSP applications. Bilinear interpolation is a basic and effective technique for resampling in computer vision and image processing. Other techniques such as linear interpolation *upsample* an audio signal to a higher sampling rate.

- **Support Vector Machine (SVM) functions**: These are a subset of ML algorithms that use **supervised learning** (**SL**), commonly used for classification or outlier detection. In this library, the SVM classification is limited to two classes, and examples are provided to simplify implementation.

- **Bayes classification functions**: Enable classification and probabilistic estimation. They are based on the powerful naïve Bayes classifiers. Although fairly simple, this technique can work well in complex real-world situations when features are independent of one another.

- **Distance functions**: Provide distances between numbers (or arrays of Boolean values) for clustering algorithms. These are key in quickly classifying data or observations into groups.

- **Quaternion functions**: These are hypercomplex, multi-dimensional numbers that have more than one complex plane. Among other benefits, they offer a helpful notation to represent spatial orientations in **three-dimensional** (**3D**) space. These functions enable common mathematical manipulations to these types of numbers, such as conjugation, normalization, and finding the inverse or product between numbers.

The CMSIS-DSP library of functions is designed to provide the essential blocks needed to create effective DSP algorithms on Cortex-M devices.

ML frameworks and libraries

CMSIS-NN, another component of CMSIS, provides optimized low-level **neural network** (**NN**) functions for Arm Cortex-M-based **central processing units** (**CPUs**). It optimizes the performance and memory footprint of NNs by offering tailored NN network kernels specifically designed for Arm Cortex-M processors and provides a significant uplift in performance and efficiency for NN functions. You can download the library from here: `https://github.com/ARM-software/CMSIS_5/tree/develop/CMSIS/NN`.

The library is categorically divided into a number of functions, briefly described next. Note that these descriptions assume a base level of ML knowledge. We will dive into ML in *Chapter 6, Leveraging Machine Learning*. This list is intended as a quick overview of the CMSIS-NN offerings:

- **Convolution functions**: These functions are used to implement customized convolution layers in **convolutional NNs (CNNs)**. Convolution results in the scalar product between an input signal and a set of weights—the single-number result is then fed into an activation function. The CMSIS-NN library supports convolutions of various sizes and requirements.

- **Activation functions**: *Activation* functions simply specify how an input is transformed into an output. This library contains several of the most common types of activation functions used in NNs: sigmoid, tanh, and **Rectified Linear Unit (ReLU)**.

- **Fully connected layer functions**: These are used to implement fully connected layers in CNNs. They provide the same mathematical result as a convolution but mandate each input node is fully connected with each output node in the layer.

- **Singular value decomposition filter (SVDF) layer functions**: A SVDF is intended to expose the underlying meaning of a matrix, represented as a linear approximation, to simplify computations.

- **Pooling functions**: Another key part of a CNN, pooling takes a large set (or *pool*) of data and summarizes it. These functions are used to identify the presence of features in a particular area and support both max and average pooling.

- **Softmax functions**: These functions are why ML probabilistic results add up to 1 (or 100%). Without these, the probability of an input belonging to certain classes could be 90% for one, 45% for another, and 20% for a third, providing useless results.

- **Basic math functions**: These offer simple, element-wide add and multiplication.

The CMSIS-NN library is tightly integrated with the TensorFlow Lite for Microcontrollers ML framework to provide optimized versions of TensorFlow Lite kernels. TensorFlow Lite for Microcontrollers is a subset of the TensorFlow ML framework from Google that is designed to run ML models on microcontrollers and constrained embedded devices with just a few **kilobytes (KB)** of memory. It is an Arm-supported frontend to deploy ML applications on Arm Cortex-M and Ethos-U-based devices. It is open sourced, has been tested extensively with the different Cortex-M processors, and is available for download here: `https://github.com/tensorflow/tflite-micro`. It can run bare-metal and doesn't require OS support on the target Cortex-M device.

We have reviewed several different CMSIS libraries that help with software startup, configuring/utilizing hardware properly, and optimizing DSP/ML workloads. The next section zooms in on two areas that warrant special consideration: security and safety software requirements.

Understanding secure software

While Arm Cortex-M processors come equipped with hardware features that provide fundamental protection for secure software, secure software needs to be written precisely to ensure that the whole system is secure. In this section, we will provide an overview of how software can be architected to benefit from the different hardware security features. We will then introduce **Platform Security Architecture** (**PSA**), a key security framework, and **Trusted Firmware-M** (**TF-M**), an open source reference implementation of PSA.

Many of the Arm Cortex-M processors have an MPU, which enables developers to configure the memory on their device into different regions with different protection levels. Programmers can leverage the MPU to protect critical processes such as the startup code and RTOS in a privileged partition, and the rest in unprivileged partitions. For example, **static random-access memory** (**SRAM**) can be defined as an execute-only memory region to prevent code injection, prohibiting user applications from accessing privileged regions used by the RTOS and thereby corrupting critical tasks. The MPU may also place communication stacks and user code in different partitions to protect the communications stack from the application code.

TrustZone for Cortex-M, which we introduced in the previous chapter, adds another layer of security that enforces hardware isolation between trusted and untrusted code. When you start your development of a project on a Cortex-M device with TrustZone, you will partition your project into a secure and user (non-secure) project. The first step is to identify all code that handles configuration and security such as the startup code, the bootloader, cryptography libraries, and firmware updates, and place them in the secure project. All the rest of the code is then placed in the non-secure project. The goal is to place the minimum amount of code in the secure project and thoroughly analyze this code for security vulnerabilities.

PSA

PSA provides a security framework for IoT devices to enable the right amount of security to be designed for devices. It is not a technology, but rather a process. PSA can be understood as four high-level stages, as outlined here:

1. **Analyze**: Understand the threat models for the current use case and define specific security requirements.

2. **Architect**: Create a design (hardware and firmware) with security requirements in mind.

3. **Implement**: Tie hardware and firmware together through open source firmware references.

4. **Certify**: Obtain certification for your product, verifying that security requirements have been met to a certain level.

Trusted Firmware for Cortex-M

TF-M is an open source reference implementation of the PSA framework for Armv8-M and Armv8.1-M architectures (Cortex-M23, Cortex-M33, and Cortex-M55). TF-M implementation is guided by a secure development life cycle workflow, which means it is secure by design. It provides PSA-compliant APIs for developers to use **Root-of-Trust (RoT)** services from the non-secure world easily. RoT is a source that can always be trusted within a cryptographic system and where security begins.

You can download TF-M from here: `https://git.trustedfirmware.org/TF-M/trusted-firmware-m.git/`.

Implementing secure software in Cortex-M devices is an often-misunderstood topic. We aim to demystify this area in *Chapter 7, Enforcing Security*.

Implementing safe software

Software development for embedded systems with safety requirements can be challenging. Safety-critical software requires special procedures during design and development. Safety analysis needs to be performed on all the software components in the project, and extensive validation needs to be done to meet the safety standards for your product. Using safety-certified software components in your application can ease your development effort.

A popular set of embedded software components that are qualified for the most safety-critical applications is Arm **Functional Safety Run-Time System (FuSa RTS)**. It applies across the automotive, industrial, and medical industries. The specific components of Arm FuSa RTS are summarized here:

- **FuSa RTX RTOS**: Deterministic RTOS that supports complex real-time applications. Enables threads, timers, memory management, and more.
- **FuSa Event Recorder**: Provides API function calls that annotate events in code that can be analyzed from memory. Simplifies testing and deployment by being non-intrusive in production code.
- **FuSa CMSIS-Core**: Vendor-agnostic software interface allowing access to the processor and device peripherals, in a safety-qualified manner.
- **FuSa C Library**: A verified subset of functions from the traditional C library for safety-critical applications.
- **Safety Package**: Holistic documentation explaining how to use FuSa RTS in a safety context.

We have now gone over some different software types and libraries addressing various computational needs and product requirements for Cortex-M-based devices.

In the next section, we will look at some use cases for microcontroller applications and explain how to select the software type and libraries for these use cases to create a complete application.

Example software stacks for common use cases

As with selecting hardware, choosing the right software stack depends greatly on your final use case. The functionality required for a low-power IoT sensor and an always-on facial recognition camera is quite different. In *Chapter 1*, *Selecting the Right Hardware*, we described six discrete areas to help select the right Cortex-M processor. In contrast, selecting the right software for a given use case, given the number of overlapping options available, has more room to include personal preference.

In this section, we will take the same use cases from the previous chapter and talk through their software requirements. We will then distill the requirements into reasonable selections for each part of the software stack. Note that there are many ways to build a working software stack for these use cases, and other options may be more optimal given previous experience, cost constraints, personal preference, development time versus quality trade-offs, and more.

Medical wearable

As a recap of the previous chapter, this product is a wrist-worn device intended to continuously monitor heart activity. It requires heightened security, low-power draw to prolong battery life, and processing power to quickly process data. The Cortex-M33 was selected as the processor to build around.

While expanding on these requirements to help select the right software stack, it is also critical to transmit the device data to a phone or hub for viewing via Bluetooth. The heart-rate monitoring functionality is critical to the device's functionality, meaning that the sensor must be deterministically monitored for data. Lastly, due to the private nature of the data, secure storage of the data is required.

Considering all these requirements, developing a bare-metal application with CMSIS is a good option for the software stack. CMSIS has middleware supporting Bluetooth drivers, can leverage TF-M to ensure secure data storage for medical information, and has device drivers to integrate sensors with ease.

Industrial flow sensor

This device measures liquids and gasses in an industrial setting, focusing on reliability and minimalism. The Cortex-M0+ was selected, being the go-to low-power/smallest-area processor.

The primary requirements for this use case are driven from the hardware perspective—highly accurate readings require highly accurate sensors. As this device is intended to be used in a closed-loop industrial setting, connected via local USB wires and not over the internet, security is not a top priority. The low-power requirement is for ease of maintenance, with minimal requirements to keep the power ultra-low via sleep states expected from the software stack. The key factor affecting the software stack is the low-cost nature of the device. Minimizing component costs leads to small memory footprints, meaning the software stack cannot afford code bloat.

In this situation, a bare-metal software stack with very few add-ons is appropriate. Examples can be taken from CMSIS to ease the software boot process, but keeping things bare-metal is ideal to minimize the code footprint to minimize the memory needed in hardware, driving down costs as much as possible.

IoT sensor

This use case is quite general and addresses the explosion of IoT sensors emerging across the world. To create a more concrete example, we can pick a stereotypical IoT sensor use case to select a software stack for. Let's say we are creating a network of 1,000 IoT devices intended to gather data across a city—humidity, temperature, noise levels, people in the area via heat signatures, and more. Each device collects a data sample every 5 seconds and transmits data back to a centralized database through Wi-Fi (readily available in the city).

To reduce complexity and data transmission costs, the devices perform computation at the edge to get a specific data point and transmit it to the database. To communicate how many people are in the area—for example—instead of sending out the raw heat-signature map every 5 seconds, only the number of people detected is sent. This requires more robust DSP capabilities. The Cortex-M7 is ideal for this use case and has enough security features to make it fit for purpose.

Taking all these requirements into consideration, selecting an RTOS for ease of development and peripheral/sensor integration makes sense. Any of the RTOSs listed previously would work well. Cloud connection and over-the-air firmware updates are necessary, and Keil RTX5 is an excellent option. Leveraging the CMSIS-DSP library will also help expedite development.

ML

This category is quite broad and hard to talk about as a monolith. Narrowing it down into distinct ML use cases will allow us to suggest sensible software stacks serving specific situations. As identified in *Chapter 1, Selecting the Right Hardware*, these distinct categories are vibration, voice, and vision.

Vibration

This includes motion detection and gesture detection. Let's consider anomaly detection in a car. The device constantly senses vibrations via an accelerometer, processes the raw data points into the ML features, and performs inference on the device to see whether the vibration is detected as an anomaly.

Translating to requirements: DSP is needed to preprocess the data into features the ML model expects, and a relatively small ML model must be run on the device directly. Assume the processor selected is a Cortex-M4, balancing DSP performance with low area for a small device.

Creating a simple bare-metal application makes sense, especially if there are not many other peripherals or connectivity requirements in the system. Leveraging CMSIS-DSP for feature extraction and tflite-micro as the ML framework is a good approach.

Voice

This includes keyword spotting and speech recognition. A common use case is a voice-enabled assistant, becoming more and more common around the home. The flow and requirements here are largely the same as the vibration use case, just replacing an accelerometer with a microphone and likely cloud connectivity. These changes require significantly more DSP and ML computation, meaning a more powerful processor such as the Cortex-M7 or Cortex-M55.

On the software side, this translates to leveraging an RTOS to ease the integration of DSP, cloud connectivity, ML, and possibly other peripherals. FreeRTOS excels in these voice use cases, containing support for connecting to the AWS cloud and the **Alexa Voice Service (AVS)** for voice control.

Vision

Includes image classification and object detection/recognition. A well-known example is a video doorbell, which can recognize faces and unlock the door for select individuals. These use cases are the most ML-intensive at the edge and require huge amounts of processing power to run facial recognition inferences.

This type of application is almost the sum of all the previous requirements put together: DSP power, ML processing, enhanced security, cloud connection via Wi-Fi, over-the-air updates, and more. At this level, an RTOS is required and should be selected based on the specific range of features you need to be successful.

Now that we have mapped the different software types and libraries you can leverage for development on your Cortex-M device to your use case, we will introduce SDKs—what they are, their main features, and how you can greatly simplify your software development by using them.

Introducing SDKs for Cortex-M

SDKs are collections of tools and software needed to get started quickly with a particular Cortex-M development board. SDKs come from a variety of sources including microcontroller vendors, board vendors, commercial or **open source software (OSS)** providers, **cloud service providers (CSPs)**, and even CPU suppliers such as Arm.

The SDK concept is not exclusive to microcontroller development and can take many forms. The most common form of an SDK is a collection of software in the form of libraries, source code, and APIs. SDKs often include examples to help developers get started or learn how to use a service or hardware feature.

SDKs may or may not include software development tools. The core tools of a microcontroller software developer are the text editor, compiler, and debugger. These tools enable developers to navigate the inner loop of code, compile, run, and debug. Engineers have individual preferences for how they would like to mix a bundle of tools and useful software. Some prefer to have everything all in one place as a single bundle, while others prefer to separate tools from embedded software and select exactly what they want. Some SDKs are even configurable for a user to select the tools and software they want to bundle, and then a custom installer is created for download and installation.

The next few pages will discuss what makes SDKs useful, what they contain, and how to select the right one for you.

Purpose of an SDK

A common demonstration of an SDK is **Blinky**, a program to turn on a **light-emitting diode (LED)**. This is the *hello world* of embedded programming. Without an SDK and no example software, it would take some effort to initialize the system, configure the hardware, and get to a place where everything is ready to turn on an LED.

With an SDK providing the underlying code to control the LED, it could be as simple as an infinite loop turning the LED on and off, as demonstrated by the following code snippet:

```c
unsigned int led_state = 1;


for (;;) {
    if (timeout_delay_is_elapsed()) {
        led_state ^= 1;
        set_led_port(led_state);
    }
}
```

With an SDK, we don't need to know anything about setting up the CPU, initializing the hardware, or even the system address for the LED. We don't even have to know how the LED is connected to the system.

Language bindings

The traditional embedded programming languages of C and assembly are now joined by new languages such as Python and Rust.

SDKs often provide easy-to-use packages for Python. With a Python microcontroller SDK, Blinky is even easier. Python SDKs for microcontrollers have become popular for education as they teach a popular and easy-to-use programming language joined with the hardware details of a microcontroller board. A simple example for blinking an LED in Python is provided here:

```python
while True:
    led.value = True
```

```
time.sleep(0.5)
led.value = False
time.sleep(0.5)
```

Middleware

SDKs typically provide a collection of middleware to make programming easier. As covered earlier in the chapter, middleware makes it easy to implement protocols such as Ethernet, USB, filesystems, storage, and **graphical user interfaces** (**GUIs**). Middleware is generally the software that is in the middle, between the hardware drivers and the application code.

In an embedded system with an RTOS or bare-metal operating environment, middleware covers many of the things we would take for granted with an OS such as Linux, and realize that these standards are generally the same for every system: not useful to recreate but important and needed. Middleware is a common component in SDKs.

Cloud connectors

IoT devices typically send data to a cloud service for analysis. Sometimes called IoT connectors or cloud connectors, these libraries are used to send data to **Amazon Web Services** (**AWS**) IoT services, to the Microsoft Azure IoT cloud, or to another cloud service. This relieves the developer from having to learn the details of sending the data so that they can just focus on what to send instead of how to send it. CSPs typically provide an SDK with libraries to connect an application directly to the cloud.

Security and encryption

SDKs may also include software for **Secure Sockets Layer** (**SSL**) and **Transport Layer Security** (**TLS**) for client-server networking applications. Using software from an SDK to create an encrypted connection between a microcontroller and a server is critical. SDKs provide the benefit of a well-tested library, which has been verified to work on a particular microcontroller and provide the best performance utilizing the hardware resources available.

An example of this is Mbed TLS, which is included in many common microcontroller SDKs. You can read more about Mbed TLS here: https://www.trustedfirmware.org/projects/mbed-tls/.

Development tools

SDKs often contain a collection of tools such as a compiler and a debugger to work with the target boards. Development tools can be command-line tools or include **user interfaces** (**UIs**). We will cover development tools in more detail in *Chapter 3, Selecting the Right Tools*, but there are a variety of options available for Cortex-M tools ranging from open source to commercial tools, each with different support options.

Each SDK makes choices about which tools to provide directly or indirectly. An SDK may bundle some tools directly into the installation to make it easy for users to start with a single download and install. SDKs may also choose to support particular tools, such as a compiler, by making sure the compiler works with the source code and libraries and providing project and build files, but users are directed to download and install the compiler from somewhere else.

Build systems

The build system is a key aspect of an embedded software project. For projects using C/C++ and assembly, it determines how the software will be compiled into an executable. There are numerous ways to build software, with so many options available that it may rival the number of programming languages available!

Build systems can be driven from a UI and have the concept of *project files*, which can be created from a UI, and easy-to-use buttons on the UI invoke build and clean operations. An example of this is the Keil µVision IDE, offering a simple interface from which to build projects. We will use this flow in several examples throughout the book.

Some build systems use text files, the command line, and scripts for complete automation. The most basic way to create a build system is using make or cmake or even just commands in a text file to run.

Some build systems are flexible and can be invoked using either the UI or the command line to provide an easy interface to get the initial project working and later the most automation to build the software.

Lower-level considerations

There can be many similarities and differences between SDKs, making it hard to narrow in on the right one for your project. The next few subsections briefly touch on specific areas to keep in mind when selecting an SDK to use.

Software examples

One of the most helpful components that an SDK can offer is software examples. Useful examples can be one of the most important time-saving features of an SDK. Make sure to use them as intended—as examples. Software examples often demonstrate the concept of how to do something but leave the details of testing and performance optimization to the developer. Don't assume too much about examples and take them as perfect; they are largely only meant to demonstrate a concept or serve as a quick start.

Software reuse

One of the challenges to consider when reviewing project options is how to reuse existing software in an SDK. Many of the RTOS and middleware components covered in this chapter may be immediately available in an SDK, but not always. It's important to consider how to integrate libraries and middleware that are not included in an SDK. While all libraries and middleware provide instructions to build it with a compiler, the details may be different from the build system and compiler used in your project. It's important to investigate these trade-offs when making SDK decisions.

For software reuse, check the following items:

- List of Cortex-M processors supported
- Compilers supported
- Build system used and complexity of migrating to another build system

Sometimes, the best way to analyze these trade-offs is to just get the software and try it out.

IDE

The IDE is a somewhat contentious topic for SDKs: developers have strong preferences for tools, especially for editing. Should it be included or not? The IDE typically merges the editor, compiler, and debugger into a single tool to make things easy and consistent. SDKs often reuse existing components to avoid building them from scratch and to take advantage of environments developers already know. Previously, desktop IDEs such as Eclipse were popular to add to SDKs. Today, cloud IDEs are becoming popular and tools such as **Visual Studio Code (VS Code)**, which can run in the cloud and on the desktop, are popular. We will leverage both desktop IDEs and cloud IDEs through the examples later in this book.

Debugging tools

Another consideration for SDKs is debuggers and hardware debug probes. Probes are used to connect a software debugger to a microcontroller development board. SDKs often include debuggers, which can trigger dependencies on hardware probes. Take a look at the combinations of debuggers, debug probes, and development boards for your project. There are many standards, but the matrix of options can get confusing. We will look at debugging tools in more detail in *Chapter 3*, *Selecting the Right Tools*.

Host platform support

In embedded software development, the computer used to code, compile, and run a debugger is called the host or host computer. The Cortex-M microcontroller running the software application is called the target. Unlike laptops or servers, microcontrollers don't have the compute resources to be both host and target, so these are separated. This division is important for SDK selection as tools used in microcontroller development typically have a limited number of host computers. Practically, this means Windows, Linux, and macOS. Engineers in a development project may have preferences for the host computer they use. More projects are also taking advantage of cloud computing to automate software build and test, so consider the OSs used in the cloud also.

Installing SDKs

SDKs have a variety of installation methods. As microcontroller projects adopt more automation, it's important to make sure tool and library installations can be easily automated. Automated build systems are moving to **virtual machines** (**VMs**) and container images, and making sure SDKs accommodate this automation is important. Watch out for tools that require a graphical installer and especially a click-through **end-user license agreement** (**EULA**) to install.

Long-term support

Many SDKs are free to use. Some offer paid support or an active subscription for support. SDK length of support is a secondary factor to review before making selections. IoT devices themselves tend to be used for a long time. Even consumer devices last far longer than mobile phones and laptops, and customers expect software updates and security fixes. Therefore, make sure the SDKs you select match up with the expected life of a product.

Looking at some available SDKs

To give you a glimpse of the wide variety of SDKs available, let's discuss a few of them. One thing you will pick up quickly is that it is easy to tell how many microcontrollers are available from a vendor by the initial look at the SDK. Vendors who are new to the market, such as the **Raspberry Pi 2040**, look very different from **NXP** or **ST** who have hundreds of different devices and are trying to support them with a common set of tools and software.

Raspberry Pi

Let's start with the Raspberry Pi Pico. The Pico, released in January 2021, is a new microcontroller at the time of writing this book. One of the unique things about the Pico SDK is that it was designed to be run on a Raspberry Pi running Linux as opposed to a Windows personal computer. All of the content of the SDK is found on GitHub. There is the SDK itself and examples. The SDK relies on numerous open source tools for compilation and debugging. The build system is cmake. Execute the following commands to download the official Pico setup script and run the installation process:

```
$ mkdir ~/raspberry-pi-pico
$ cd ~/raspberry-pi-pico
$ wget https://raw.githubusercontent.com/raspberrypi/pico-
setup/master/pico_setup.sh
$ chmod +x pico_setup.sh
$ ./pico_setup.sh
```

The Raspberry Pi Pico SDK can run on Windows and macOS, but it works great on Linux.

NXP

NXP provides the MCUXpresso SDK to simplify your software development on NXP boards with Arm Cortex-M CPUs in them. It is free and available for download directly from the NXP website. When you download the SDK, you are prompted for the target board you plan to use it for. You can customize your SDK download package to include the components you will need for that board. The components include libraries such as CMSIS-DSP, middleware, and RTOS support. For example, if you download the SDK for use with the **LPC55S69-EVK** board, which has an Arm Cortex-M33 CPU, you might want to add support for the TF-M library. You can also select the host OS and compiler toolchain you would like to use. Note that Windows, Linux, and Mac are all supported.

ST

ST also provides a whole suite of software development tools for its Arm Cortex-M-based microcontrollers. These tools are referred to as STM32Cube tools and provide a range of support for their portfolio of boards. This includes a graphical tool for generating initialization code for your device, as well as project management and debug features. As with other SDKs, middleware components for peripherals and connectivity modules on the ST boards are also provided. You can even download the STM32Cube package for the specific **microcontroller unit** (**MCU**) family you are targeting. The tailored packages contain several out-of-the-box examples that are quite useful to get started on your board right away.

These tools were previously supported only on Windows but recently have added support for Linux as well.

Summary

In this chapter, we first looked at the two main software types ported to Cortex-M microcontrollers— bare-metal and RTOS. We examined several different software libraries specialized for DSP, ML, and security applications.

We then walked through a few different use cases for microcontroller applications and highlighted the software stack most applicable for that use case. Finally, we introduced what SDKs are, their main features, and how they can help accelerate your Cortex-M software development.

To sum up, we have looked at a wide variety of software available for Cortex-M microcontroller applications. Thankfully, there is a large ecosystem of both commercial software and OSS available to use, which makes development easier and helps to create performance-optimized applications.

Next up is *Chapter 3*, *Selecting the Right Tools*, the last chapter of this part, where we will navigate through the assortment of tools and environments that enable Cortex-M software development while understanding the benefits of different options.

Further reading

For more information, refer to the following resources:

- PSA Certified resources: `https://www.psacertified.org/`
- TF-M documentation: `https://tf-m-user-guide.trustedfirmware.org/`
- CMSIS documentation: `https://arm-software.github.io/CMSIS_5/General/html/index.html`
- Cmake build system documentation: `https://cmake.org/`

3
Selecting the Right Tools

The Cortex-M development tool ecosystem offers a wide variety of options for software developers. In the previous chapter, we spent time reviewing specific **software development kits** (SDKs), some of which contain development tools. In this chapter, we will look specifically into Cortex-M software development tools. The ultimate end goal for most microcontroller projects is to deliver high-quality, secure software applications, optimized for variables such as performance, power, or cost. Selecting helpful development tools is essential in reaching this final state.

Similar to the first two chapters, this chapter is intended as an overview of which tools are available to Cortex-M software developers. The benefits and drawbacks of different tools are presented for you to make an educated selection about the right set of tools for your specific project.

We start by first explaining the options available to run Cortex-M software during the development process, as there are a number of ways to prepare and test the software before running it on the final hardware. Next, the traditional compiler, debuggers, **integrated development environment** (IDEs), and performance analysis tools are explored. Finally, we look at how engineers build up a collection of tools to operate efficiently on a day-to-day basis.

To sum up, we'll be exploring the following areas in this chapter:

- Development platforms
- Compilers
- IDEs
- Development environments

It's important to remember that development tools are typically a means to get to the final product. The usage of various tools may come and go during different phases of a project, and not every tool is applicable to every project. We encourage you to keep an open mind, learn about the wide range of technologies that apply to Cortex-M software development, and focus on ways to improve the important factors of quality, security, and **time to market** (TTM) for your projects.

Another way to think of development tools is in the context of the inner loop of code: compile, run, and debug. These are the daily activities of an active software development project that have the most impact on project results. These basic steps help frame the context of development tools.

Let's get started by looking at various types of development platforms available for software engineers.

Examining development platforms

Development platforms generally refer to the ways we *run* or execute software. Of course, the final destination to execute software is on the physical product being created, but during development, this hardware system may not be available. If the hardware is not available, software developers turn to one or more development platforms to test what they can and do as much as possible while waiting for the final hardware. There are a number of ways to perform early software development, each with its pros and cons. In this section, we will highlight some alternatives and provide guidance on why or when to use them. Cortex-M developers are most likely to focus on the first set of platforms, which we call *software-centric*. However, some developers may also use *hardware-centric* platforms in large teams developing low-level firmware for chips. We will discuss both sets to provide a full context of what is available.

Software-centric development platforms

Let's look at the most common development platforms that software engineers turn to while developing software before hardware is available. Note that some of these platforms are complementary to your finished device and may be useful for debugging or automated testing even when you have production-ready hardware in hand. It may make sense to use multiple platforms at once for different reasons, so keep an open mind to what makes sense for your particular project.

Development boards

The most common Cortex-M development platform is a development board. Their ubiquity is why we dedicated a section of *Chapter 1, Selecting the Right Hardware*, to how to select the right one for your project. We will also use development boards in future chapters to demonstrate various concepts of Cortex-M software development.

The easiest way to get started with Cortex-M software development is to get a board and find examples to run on the board. Many times, precompiled examples are available that can be run without installing any tools. But to really get started, developers will set up tools and—potentially—an SDK.

If an off-the-shelf microcontroller is being used for the project, there are often development boards with the same microcontroller available. This makes it easy to get started right away with software development. If your project is not using an off-the-shelf microcontroller but instead designing custom silicon, then most software developers will find a development board with the same Cortex-M processor and a similar set of peripherals before getting started with software development. Development for custom silicon will often shift to some of the other platforms outlined in this section.

Development boards are popular because they run at the same or similar speeds to the final product, are easy to obtain, and have good connections to tools such as debuggers.

Another topic around development boards is automated testing. While it's possible to create a lab full of development boards for automated testing, the process can be difficult to scale for large numbers of boards. The maintenance cost typically far outweighs the sum of the purchased boards as a non-trivial amount of people-hours are spent maintaining the board farm. Remote control and the ability to connect to **input/output (I/O)** interfaces can also be challenging. We will look at some alternatives to building your own board farm for automated testing in *Chapter 9, Implementing Continuous Integration*.

Virtual platforms

A virtual platform is one name for a simulation model that executes the instruction set of the Cortex-M microcontroller. After development boards, virtual platforms are the most common way to execute software and debug it. The primary benefit of virtual platforms is that no physical hardware is necessary. No need to order boards, find cables, use debug probes, or anything physical. This makes starting software development with virtual platforms fast and easy.

Virtual platforms use instruction translation, also called code translation, to translate Arm instructions into the instruction set of the host computer. This often means translation to x64 instructions for Intel or **Advanced Micro Devices (AMD)** processors running on a Windows PC or Linux server. The translated code blocks are then cached so that when the code is executed again, the translation is already available and executes faster. More recently, Windows and Linux machines have appeared with Arm processors, so the instruction translation is simplified.

Virtual platforms provide a *programmer's view* of the hardware being simulated. This indicates that the model provides a full description of the system memory map and programmable registers that are visible from executing software. A good programmer's view model makes it impossible to distinguish between a virtual platform and the physical hardware.

What programmer's view models do not provide is detailed performance information or execution timing. There are no **central processing unit (CPU)** microarchitecture details such as CPU pipeline stages, delays accessing caches or memory, and most other timing information. Note that this makes it impossible to use virtual platforms for detailed performance analysis. Virtual platforms can be used to count the number of instructions executed and use this count for directional improvements in performance. We will provide an example of this behavior in *Chapter 5, Optimizing Performance*.

Virtual platforms use additional optimization techniques to provide fast execution performance. The most time-consuming activity of a microprocessor is fetching instructions and reading/writing memory data. To increase virtual platform performance, the memory-access activity is optimized by streamlining parts of the simulation. Put simply, it is able to run faster by simulating less. Using various techniques, all the details of the protocol between the processor and memory are completely skipped.

These optimizations lead to fast simulation speeds. Virtual platforms can run between 50-5,000 **million instructions per second** (**MIPS**) on a typical workstation. In real terms, this enables a **real-time operating system** (**RTOS**) to boot instantly, and software—in general—can run many times faster than the physical Cortex-M-based development board.

There are two types of virtual platforms, as outlined here:

1. The first type models an existing physical board. An example of this is the **Cortex-M Microcontroller Prototyping System** (**MPS**) development board. This is a board with a **field-programmable gate array** (**FPGA**), memories, and other hardware, which provides a way for software developers to quickly get started with Cortex-M software development. Because it supports different FPGA images, it can be reprogrammed to run a variety of Cortex-M processors for comparison and benchmarking without purchasing multiple boards. Virtual platforms of the MPS2 and MPS3 development boards exist, and they model the FPGA's physical hardware including the **liquid-crystal display** (**LCD**), buttons, switches, and peripherals. This makes it easy to develop software on the virtual platform and move it directly to the physical board for performance analysis. Ideally, these virtual platforms of physical boards are identical and run the exact same binary software. Practically, this isn't always possible, so some modifications may be needed for the software to work on the virtual platform and the physical board.

2. The second type of virtual platform is a customized platform created to model a future hardware system. Virtual platforms are commonly used for early software development for a new silicon design. Companies doing silicon design have a team of virtual platform creators who can make custom virtual platforms that match the hardware design being created. This enables software teams to start very early and be ready when silicon and development boards become available. Custom virtual platform creation requires a mix of system design skills, hardware understanding, and software skills to create a virtual platform for software engineers to use. Virtual platform creation is a unique career path for those engineers who enjoy both hardware and software topics. There are a number of languages, including SystemC, that can be used for modeling peripherals and other system functions. These custom models are combined with a model library of processors and other standard peripherals, then assembled to create a custom virtual platform.

There are open source and commercial virtual platform tools available for Arm processors, including Cortex-M processors. These can be used to immediately run already created virtual platforms and create custom virtual platforms. Here is a list of virtual platform technologies:

*   **Arm Fast Models**: Fast Models are accurate, flexible programmer's view models of **Arm Internet Protocol** (**Arm IP**). They allow you to develop software such as drivers, firmware, **operating systems** (**OSs**), and applications prior to silicon availability. You can find more information here: `https://developer.arm.com/tools-and-software/simulation-models/fast-models`.

- **Imperas and Open Virtual Platforms**: Imperas has a range of virtual-platform creation tools and model libraries for a variety of CPU architectures. **Open Virtual Platforms** (**OVP**) was created by Imperas to simplify the creation and execution of models. You can review their home pages for further context at `https://www.imperas.com/` and `https://www.ovpworld.org/`.

- **Renode**: Renode is an **open source software** (**OSS**) development framework with commercial support from Antmicro that allows for the developing, debugging, and testing of multi-node device systems in a manner that is reliable, scalable, and effective. More information is available at their website: `https://renode.io/`.

- **Quick Emulator** (**QEMU**): A generic and open source machine emulator and virtualizer that includes models of multiple Arm systems. Learn more and access the platform here: `https://www.qemu.org/`.

Host code development

Embedded software engineers have always done some level of early software development using the native instruction set of their computers. This means developing part of an application intended for a Cortex-M-based device on a traditionally Intel or AMD-based PC. For compiled languages such as C and C++, the software applications were likely built and run on a different CPU architecture than the final hardware. However, for functional testing, this can be useful to speed development using native tools instead of microcontroller tools. This is covered further in *Chapter 9, Implementing Continuous Integration*.

The Arm architecture is now being used more and more in laptops and servers. This makes it possible to perform early microcontroller software development on the same Arm architecture as the microcontroller.

RTOSs also offer the ability to build and run on a host computer and use debug and analysis tools from the same machine to debug and profile.

There are some significant differences between the architectures used on a server system such as **Amazon Web Services Elastic Compute Cloud** (**AWS EC2**) instances based on AWS Graviton processors. Amazingly, however, many of the same microcontroller instructions can run directly on the Arm Neoverse processors used in cloud servers offered by AWS, Oracle, and Google.

Cloud computing

Cloud computing is a trend across all types of software development, and microcontrollers are also now following this trend. More work is being done on remote machines, often in a public cloud such as AWS. Just as with general-purpose computing, software development platforms in the cloud make it easy to quickly start and stop development platforms.

Virtual platforms are now being offered as a cloud service to easily start, connect, and stop without any downloading and local software installation. Two other cloud solutions are a virtual **internet of things (IoT)** board and a physical IoT board in the cloud. These solutions benefit from the *pay-for-what-you-use* model and provide access to a wider variety of development boards without purchasing them.

Virtual IoT board

Virtual representations of physical boards can be created using a hypervisor on an Arm server to model a Cortex-M development board. The hypervisor serves the same purpose as a general-purpose hypervisor, whereby cloud users share the hardware and are isolated from each other by the hypervisor. The same concept is now available where the compute instance is a virtual IoT board. By running on the Arm architecture and providing built-in modeling of peripherals, it creates a solution that is faster than virtual-platform models based on instruction translation. Virtual IoT boards are accessed directly in a browser, using an **application programming interface (API)**, or via a **virtual private network (VPN)** to connect to the virtual board.

Arm Virtual Hardware (AVH) currently offers a number of Cortex-M and Cortex-A virtual boards that can be used for software development and automated testing. The best way to think of virtual IoT boards is like a **virtual machine (VM)** in the cloud, but, in this case, the VM is an IoT board.

Remote access to physical IoT boards

Another cloud-like access model is to make physical boards available via the internet. Just as we access remote machines in the public cloud and services from AWS, a similar concept is possible for Cortex-M development boards. Users can log in to a web console, connect to remote physical Cortex-M development boards, load software, and debug applications. As with the virtual IoT board, the entire process can be scripted and controlled by an API.

Providing development platforms as a service is an emerging space, so we encourage you to keep an eye on new developments. Now, let's shift to hardware-centric development platforms and survey some of the tools used in custom silicon development.

Hardware-centric development platforms

While software engineers are typically looking for early access to the hardware design, hardware engineers are typically looking to gain confidence that the software will run on the hardware being designed. Not all projects will involve custom silicon design, but there is a trend for projects to investigate the possibility and benefits of creating custom silicon for high-volume applications as they deliver additional optimizations compared to using existing silicon.

While in the future using these types of techniques may make sense if your company decides to develop custom silicon, right now few Cortex-M software developers have a need for these types of platforms. If you are targeting an existing microcontroller with a Cortex-M processor, these options are only here for your awareness, not as reasonable options for development!

With that said, we will dive into some specifics for the few developers targeting custom Cortex-M silicon over the next few paragraphs. Custom silicon designs are created using a **hardware description language (HDL)**. The process involves describing the hardware using an HDL such as Verilog. Once the hardware is described, it can be executed using a number of platforms or execution engines. We will simplify the discussion to three primary techniques, as follows:

- **Logic simulation**: A logic simulation refers to an event-driven logic simulator running on a general-purpose workstation or server. Simulators read the hardware design description and use advanced compilation techniques to turn this hardware description into an executable that can be run. When the simulator is run, it keeps track of the binary values of each signal in the hardware design and propagates binary values throughout the design description.

- **Emulation**: Emulation is the process of mapping the hardware design description into a dedicated hardware platform designed to increase performance. This purpose-built hardware machine typically leverages custom-designed processors or an array of FPGA chips to execute software on. Emulators increase performance by building parallelism into the hardware processing, far more parallelism than what is possible on a simulator running on a multi-core workstation. Emulators can handle very large designs and are often used for executing an OS and testing applications for high-end products such as the next generation of servers or mobile phones.

- **FPGA prototyping**: FPGA prototypes describe the construction of a custom hardware system or the use of a generic array of FPGAs to build a prototype. The prototype idea here is that a representative hardware system can be created in less time than the final silicon device by giving up requirements such as performance or packaging. For example, the FPGA prototype will be slower and larger than the final custom silicon product, but having it sooner is worth the trade-off to execute software and confirm the hardware functions as expected.

There are numerous other products based on mixing any of the preceding technologies. For example, emulation vendors will join Arm Fast Models running on a workstation to an emulator to obtain increased performance in exchange for less detail in the hardware system.

Hardware-centric development platforms are provided by companies in the **electronic design automation (EDA)** industry. EDA is a complete industry by itself and provides the needed products to create silicon devices. Cadence Design Systems, Siemens EDA (formerly Mentor Graphics), and Synopsys are the top three EDA companies providing the hardware-centric development platforms listed previously.

It can be hard to know which of these development platforms to use in a particular project. Project teams make choices to adopt the ones most useful to cover project needs and ignore the others. To give some guidance related to the alternatives, a short set of metrics is described in the next section. Refer to them when you need to decide which development platforms to evaluate for your next project.

Metrics for evaluating development platforms

There are eight distinct metrics that we use to differentiate development platforms. We will go over each metric to help you determine if it needs to be prioritized in your next project. An overview of the metrics is provided in the following screenshot:

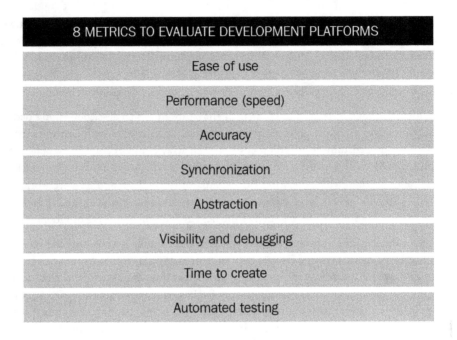

Figure 3.1 – Evaluating development platforms

Ease of use is a good first metric. Some development platforms require a complex setup; both software and special hardware may be required. Some development platforms are very easy to use and operate entirely in a browser with no software to install and no required hardware. They can work on any kind of computer and don't have dependencies on an OS such as Linux or Windows. The more complex a development platform, the more support will be required to help platform users. Ultimately, this is a metric unique to each person; trying to run a software example on a few different development platforms will quickly measure what your tolerance is for ease of use.

Performance (speed) relates to how fast the development platform is running, usually in comparison to the final product being created. Cortex-M software developers are looking for performance as close as possible to the actual product. There is a relationship between the performance of a development platform and the details of the platform. Some technologies get slower as the design size or complexity increases, and other technologies are independent of size or complexity. Designs using Cortex-M processors often focus on lower power and don't have clock frequencies as high as general-purpose computers, so it may be possible that a development platform is faster than the actual product being designed.

Accuracy describes how close the development platform is to the final product. A virtual IoT board may look very much like the real thing, but there is modeling involved and all details may not be exactly the same. For any development platform, it's important to know about any areas where abstractions are used, or details left out. Virtual platforms omit numerous details about how a hardware system works. They are useful for software development, but it's important to know what they are good for and what they are incapable of doing.

Synchronization is about how multiple time domains interact. Often, tools will use shortcuts to speed up the performance, such as the **Direct Media Interface** (**DMI**) described previously, at the expense of changing system timing. Think about a virtual prototype with two Cortex-M processors. A virtual prototype will run groups of instructions on each processor model serially and then let the other processor take a turn. In a real microprocessor, the two Cortex-M processors will execute in parallel. This synchronization impacts software timing and even functional results.

Abstraction explains why development platforms have limits on the types of software that can be run or the results that can be obtained. A logic simulator can be very accurate but may be limited to running benchmarks up to a given number of clock cycles, or the runtime becomes days long. Virtual platforms may not be able to run certain software that requires features not modeled by the development platform. An example is custom instructions that are added to a Cortex-M processor. If the standard model doesn't know the custom instructions, the software cannot run on the model.

Visibility and debugging provide the means to understand what happens when something goes wrong. Development platforms offer a wide range of techniques to debug interactively or trace execution and events without stopping the execution. Depending on the software development task, more or less visibility and debugging may be required. Sometimes, it's not possible to stop and start a system or query information about memory or register values.

Models always take *time to create*. It's important to measure how long a development platform takes to create and whether it's possible to create a meaningful solution in the timeframe of the project. Remember—development tools are a means to an end, and if it takes longer to create the development tool than to complete the entire project, the tools are not going to be useful. Time to create may also refer to the time to construct a lab with the needed hardware boards and test equipment.

Performing *automated testing* means scripting and automating software tests. During testing, it's common to vary a number of parameters, such as the compiler or processor used. Projects delivering software libraries usually want to test the functionality and performance of the library across a wide variety of hardware since users may try the library on many different systems. Automating the testing and running across a variety of parameters saves significant time.

We have covered a wide range of development platforms, both software-centric and hardware-centric. All of these are not possible on a single project, but it's important to understand what is available, the pros and cons, and how to make good choices when thinking about which of them to use on your next project.

Next, we will walk through the primary compilers available in the market to build your code targeting Arm Cortex-M devices. After that, we will outline the different IDEs and their features that you can utilize to simplify and accelerate your software development.

Exploring compilers

In the general software development industry, the term *compiler* may mean slightly different things to different people. A *toolchain* typically includes a compiler, an assembler, a linker, and a disassembler. Although not totally correct, the collection of tools may also be called a compiler. With these utilities, the source code that you write for your application, along with any software libraries you have integrated, gets converted into the final executable that runs on the Cortex-M device. As we walk through some of the toolchains available for your Cortex-M project and their features, we will also highlight the utilities in each of these toolchains that perform these functions.

Several factors can influence your compiler choice. It depends on what your target Arm processor is and whether you are more inclined to use an open source compiler (available at zero cost) or a commercial compiler (costs money). Other factors such as support from the vendor, code size, and performance of the compiler can also impact your decision.

Arm Compiler for Embedded

Arm Compiler for Embedded, formally known as **Arm Compiler 6 (AC6)**, combines components of the **Low-Level Virtual Machine (LLVM)** Compiler infrastructure with an Arm linker and libraries. A major advantage is that it is co-developed and co-verified with the development of Arm processor cores. As a result, support for the newest Arm processors is available in Arm Compiler for Embedded before other compilers.

It has the latest support for C++11/14. It is supplied with the MicroLib C library, which minimizes code size. Features such as link-time optimization use various techniques to dramatically reduce your code size. Compiler diagnostics are often overlooked because they are generally very basic and lack actionable information. Comprehensive and actionable diagnostics are provided by Arm Compiler for Embedded.

Arm Compiler for Embedded fully supports all cores in the Arm Cortex-M family, including the newest Armv8-M architecture cores (Cortex-M55 and Cortex-M85). With Arm Compiler for Embedded, you can also build secure applications based on TrustZone.

The main components of Arm Compiler for Embedded are set out here:

- `armclang`: The compiler and assembler, which translates source code into binary object files
- `armlink`: The linker, which gathers the compiled object files and combines them into a single executable application
- `fromelf`: The image conversion utility, which can generate textual information about your compiled application such as disassembly, code, and data size

Arm Compiler for Embedded is built into IDEs offered by Arm—**Keil Microcontroller Development Kit (Keil MDK)** and Arm Development Studio—which we will cover in the next section. It is commercially licensable and available for purchase here: `https://developer.arm.com/tools-and-software/embedded/arm-compiler/buy`.

Arm Compiler for Embedded FuSa

Arm Compiler for Embedded FuSa is a qualified toolchain optimized for running safety-critical applications on Arm processors. It is a branch of Arm Compiler for Embedded that has been qualified for development in safety-critical applications. It is certified by the TÜV SÜD standards body for safety applications up to the highest levels. The toolchain is commercially licensable here: `https://store.developer.arm.com/store/embedded-iot-software-tools/arm-compiler-6-functional-safety`.

GNU Arm Embedded Toolchain

The **GNU Arm Embedded Toolchain** is an open sourced **GNU's Not Unix** (**GNU**) toolchain primarily focused on systems running bare-metal code or a simple RTOS, targeting Arm Cortex-M series processors along with other architectures. It contains integrated packages with the **GNU Compiler Collection** (**GCC**), including the compiler, linker, and libraries you will need to run code on your Arm Cortex-M device. A benefit of GCC is that it is completely free to use. GCC also has large community support, and while it doesn't have a professional support model like commercial compilers do, bug fixes and patches are regularly made available.

The main components of this toolchain are listed here:

- `arm-none-eabi-gcc`: The compiler
- `arm-none-eabi-as`: The assembler
- `arm-none-eabi-ld`: The linker
- `arm-none-eabi-objdump`: The disassembler

This toolchain is command-line-only and doesn't come with an IDE. It does come with a debugger: `arm-none-eabi-gdb`. The *gdb* in the utility name stands for GNU Debugger and can be used to debug your source code by setting breakpoints, stepping through the code, viewing stack traces, and so on.

You can download the toolchain from here: `https://developer.arm.com/tools-and-software/open-source-software/developer-tools/gnu-toolchain/downloads`.

Now that we have covered some common compilers available for the software development of your Cortex-M project, let's move on to the IDEs.

Navigating IDEs

Arm Cortex-M processors are supported by all the major IDE vendors. Some of the IDEs are Eclipse-based, some are based on a proprietary **graphical user interface** (**GUI**), and some are open source. They are available at a variety of price points and licensing business models, ranging from free to low-cost based on open source to higher-priced proprietary IDEs.

As we introduced earlier, IDEs are a complete solution for code development and debugging. The distinction between SDKs and IDEs is subtle, but differentiating between them is helpful in thinking about different groups of tools. Some (but not all) SDKs include IDEs, such as the NXP MCUXpresso SDK, which includes software libraries for your particular NXP board and also an IDE to develop on. Some SDKs do not include IDEs—for example, the Raspberry Pi Pico SDK. One definition of an IDE is a tool with a GUI that is used for editing, compiling, and debugging. The **user interface** (**UI**) takes care of calling the individual tools as needed, so users do not need to learn the details of invoking each tool.

All major IDEs support Windows and a few support Linux. While all IDEs largely perform the same tasks, the one you ultimately select typically comes down to your and your team's personal preference. In some cases, IDEs are purchased to access the compiler toolchain bundled with it.

In this section, we will provide an overview of some of the most popular IDEs for your Cortex-M software development. We will also leave you with links to other IDEs for reference.

Keil MDK

Keil MDK is one of the popular and comprehensive software development solutions for Arm-based microcontrollers and includes all components that you need to create, build, and debug embedded applications. The MDK product is broadly composed of tools and software packs.

Keil MDK includes two components: MDK-Core and Arm Compiler for Embedded.

MDK-Core is a µVision-based IDE with support for all Cortex-M devices including the new Armv8-M architecture. With the µVision IDE, you can create individual projects or multiple project workspaces, and interface to third-party tools for version control. It also contains µVision Debugger, with which you can debug your application code using typical debug features such as breakpoints and watchpoints, and debug device peripherals with full visibility.

MDK includes support for Arm Compiler for Embedded with assembler, linker, and highly optimized runtime libraries that are tailored for optimum code size and performance.

Software packs are the second component of MDK. More specifically, support and IDE integration of **Cortex Microcontroller Software Interface Standard** (**CMSIS**) software packs are included with MDK. With software packs, you can add device support and software components that you can use as building blocks for your application. They may be added at any time to MDK-Core, making new device support and middleware updates independent from the toolchain. Software packs include CMSIS libraries, middleware, board support, code templates, and example projects.

MDK-Middleware provides tightly coupled software components that are specifically designed for communication peripherals in microcontrollers. For example, it contains the IPv4/IPv6 networking communication stack to enable IoT applications.

The full MDK suite is commercially available here: `https://store.developer.arm.com/store/embedded-iot-software-tools/keil-mdk`.

Arm Development Studio

Arm Development Studio is a full-featured IDE built on Eclipse for software development on all Cortex-M processors as well as other families of Arm processors. Arm Development Studio includes an Arm debugger and contains a license to use the µVision debugger. The Arm debugger supports both symmetric and asymmetric multi-core debugging on Arm-based **systems on a chip (SoCs)**. As with Keil MDK, Arm Development Studio includes Arm Compiler for Embedded and software packs. Virtual platforms are included in Development Studio, providing a development platform for any Cortex-M processor without the need for a physical board to get started. Development Studio also includes a library of bare-metal software examples ported to several Arm platforms, including ones with a Cortex-M processor.

An advantage with an Eclipse-based IDE such as Development Studio is that you can install a whole suite of plugins on top of the base installation, so you could use things such as **Eclipse Embedded C/C++ Development Tools** (**Eclipse Embedded CDT**) and anything else of interest to you. Eclipse Embedded CDT is an open source project with a family of Eclipse plugins and tools based on the GNU toolchain that you can use for your embedded Arm development. You can view and purchase Development Studio from a variety of distributors or from Arm at `https://developer.arm.com/tools-and-software/embedded/arm-development-studio/buy`.

IAR Embedded Workbench for Arm

IAR Embedded Workbench for Arm is an IDE made by IAR Systems. It supports all Arm Cortex-M processors. The IDE includes the IAR C/C++ compiler and debugger for your Arm Cortex-M software development. You can use CMSIS-Packs and CMSIS-DSP libraries with IAR Embedded Workbench. It also includes RTOS plugins and support for communication stacks and middleware. It is a commercially licensed IDE with different product editions. The Cortex-M product edition covers support for all the Arm Cortex-M family of processors and is most suitable in this context. You can purchase IAR Embedded Workbench for Arm here: `https://www.iar.com/products/architectures/arm/iar-embedded-workbench-for-arm/`.

Arduino IDE

Arduino IDE is an open sourced IDE that is widely used for embedded development in the Arm community. It comes with out-of-the-box support for all Arduino boards, which includes boards with Arm CPUs in it. It uses the GCC Arm Embedded toolchain for software development on Arduino boards with Arm CPUs. The IDE can be used both offline and online.

Note that this differs from traditional IDEs because the code is Arduino-specific. It also imposes other limitations through its sketch file format and is generally best suited for developing on Arduino boards and related devices.

The first Arduino board based on an Arm microcontroller was the **Arduino Due**, which is easily programmed using the Arduino IDE. Since then, Arduino has added several other boards with an Arm Cortex-M CPU in it, including the Nano 33 IoT featuring a Wi-Fi module or the Nano 33 BLE Sense featuring **Bluetooth Low Energy** (**BLE**) and several environmental sensors. As with other IDEs, it comes with several features that simplify code development on your target board. One thing to note here is that Arduino boards are extensively used by hobbyists. There is a large community of support, with blogs and example projects around them.

Arduino IDE can be downloaded for free at `https://arduino.cc`.

Visual Studio Code

Visual Studio Code (**VS Code**) from Microsoft is not quite a traditional IDE. It is, however, worth mentioning in this section as it is often used by software developers as a code editor and development environment for building, debugging, and deploying software to Arm-based devices. Depending on the features and functionality you need for developing the software for your Arm Cortex-M project, whether it is the language that you are coding in, the compiler you are using, or the debug tools you need, you can install these using **VS Code extensions**. One of the newer VS Code extensions released by Microsoft earlier this year is the **Embedded Tools extension**. Using this new extension and a package manager utility called `vcpkg`, you can get your embedded tool dependencies easily and quickly debug your embedded code in VS Code. There are a few examples available that walk you through the usage of VS Code for developing and deploying software on IoT boards with an Arm **microcontroller unit** (**MCU**), such as the one here: `https://github.com/azure-rtos/getting-started/blob/master/MXChip/AZ3166/vscode.md`.

Now, these are only a few of the IDEs available for use today. Covering them all is out of scope for this book, but you can search around and find others that we have not covered here. There is one additional IDE, Keil Studio Cloud, that we will introduce through an example in *Chapter 8, Streamlining with the Cloud*.

In the next section, we will review how to use development platforms, compiler toolchains, and IDEs together in a cohesive environment.

Understanding development environments

Development environment is another overused term in the industry. It has a different connotation in website development—referring to server levels intended for development, testing, and deployment—and is often conflated with an IDE. For the purposes of this book, we define your development environment to simply be the software environment where you work, the substrate through which you develop your product. It can be best understood through examples, and we can helpfully break down different development environments into three main categories.

The first is a *local environment*, such as a PC running a standard OS (Linux, Mac, or Windows). The second is a *virtual environment*—for example, Docker or VirtualBox. The third is a *cloud environment*, such as AWS or the **Google Cloud Platform** (**GCP**). This section breaks down the key considerations when selecting one of these types of development environments.

This is a decision often made by default as we typically prefer one environment, or perhaps our company has used the same environment for years and it's taken for granted. It deserves a conscious decision; however, your development environment can fundamentally affect which tools or software you can use, as well as your team's overall productivity.

An important facet of this decision is determining how your team does (or wants to) develop software. Every software project has some version of *interactive* development, which involves an individual writing code, compiling it, running it on some platform, and debugging any errors. Some projects (mostly at the professional level) also utilize *automated* development, which automatically compiles, runs, and diagnoses any errors found in code on a nightly basis. These two different styles of development are enabled with different tools and have different requirements, and therefore often require different development environments.

This section will discuss the different needs of both *interactive* and *automated* development styles and goes through the three development environments identified—local environments, virtual environments, and cloud environments—to guide your decision-making process.

Interactive versus automated development

As previously mentioned, all software projects involve interactive development. Its four steps (coding, compiling, running, and debugging) are fundamental to developing software and are likely familiar to you already.

The following diagram illustrates the software development cycle:

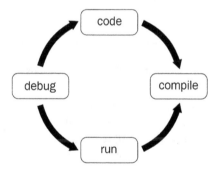

Figure 3.2 – The software development cycle

You code through an IDE, compile with a compiler, run on your selected platform, and debug either through the same IDE or possibly a command line/terminal. It is the daily routine for millions of software developers around the world.

Automated development is intended to perform automated testing on some set schedule (such as nightly) or on events (such as pushing code to a shared repository). It is very useful to validate code quality and can keep track of code progress. This process is commonly referred to as **continuous integration** (**CI**) and we dedicate a chapter to expand on the topic (*Chapter 9, Implementing Continuous Integration*). With regard to development environments, you will likely select a different environment for your interactive development versus your automated development. The reasons are concisely described in the following table:

	Interactive	Automated
Simplicity Mostly all developers know how to use the environment and tools well	**Critical** Where most developers do day-to-day work, must be as streamlined as possible	**Helpful** Typically, only one or a few dedicated members actively work on the details
Replicability The same errors can be seen in other developers' environments	**Helpful** May not be developing the same area of code, or when working alone	**Critical** To streamline development, enabling all developers to be able to see the error
Longevity Able to be used for longer than the current project	**Depends** May need to change environments to match the needs of diverse projects, but it certainly helps to have a consistent bit for years to come	**Critical** The value of scaling up tests comes when you don't have to keep setting up unique automated flows each time

Table 3.1 – Value of simplicity, replicability, and longevity to types of software development

Simplicity refers to how easy the environment is to work with for the members of the team. This is critical for interactive development as this is where development will be done every day. If you use AWS and are totally new to cloud-based environments, small issues with setup or in daily interaction will add up over the course of a project. For automated development, the bulk of the setup and maintenance of the testing flow is typically handled by select members of the team, and hence the need for a simpler environment is lessened. For lone developers, however, it makes life a lot easier to have a simple flow.

Replicability is about minimizing the number of times fellow developers state, "I cannot see the same problem you are seeing." This can be quite annoying as you have to spend time debugging your environment before you can start debugging your code together. For interactive development, it is helpful to have as much replicability as possible but this may not be essential, especially if teams focus on coding different sections at a time. For lone developers, this only matters if you want to code on different environments for the same project—for example, on a Mac at home and a Windows machine at work. In the automated development sense, it is critical to be able to replicate errors as the primary purpose of these tests is to identify errors for resolution.

Longevity considers the future—how much you can reuse your environment for the next product. For interactive development, this may or may not be a concern that largely depends on your situation. In automated development, much of the upfront time is spent building up your development environment with tools and software to streamline tests. It is, therefore, much more beneficial to pick an environment, set it up properly, and leverage it for many projects' testing needs.

With those factors in mind, we will now go through each type of environment—*local, virtual, cloud*—and assess them on their simplicity, replicability, and longevity.

Local environments

Local environments are an OS (Linux, Mac, or Windows) running on a PC or laptop. This is often the default option, given the ubiquity of personal and work laptops. Each of these OSs has pros and cons when it comes to developing software. Some specific idiosyncrasies are called out next. Note that this is a subjective topic, and the following list is influenced by our own preferences:

- **Windows**:

  - Has different line endings than Mac or Linux, meaning transferring files or developing partially on Windows versus other OSs deserves special attention.
  - Is more UI-focused. It is easier to do things graphically but more difficult to work from the **command-line interface (CLI)**.

- **Linux**:

  - Is for more advanced users. It is great for having control over your environment as it is easy to install libraries and tools.
  - Is easier to mess up your system as well.

- **Mac**:

  - Similar to Linux in some ways as it is derived from the same OS as Linux (Unix), thus sharing many properties.
  - It combines the benefits of a well designed UI with a powerful CLI.

If your project requires certain tools or software that are much more Windows-friendly, such as Keil MDK, the decision to use a local Windows development environment may be made for you. Other tools lean heavily toward Linux environments, which leaves your options open to be local, virtual, or cloud-based. It is possible to set up Windows or Mac-based virtual or cloud environments, but it is considered trickier than with Linux.

Using local environments, as this is often how we get started using PCs, is often the simplest approach for developers. Longevity also typically is not an issue. The primary issue revolves around replicability. When development teams use their personal laptops to develop code, it is very difficult to control for small variations in OS versions, software libraries, and tools, which can all lead to poor replicability of results.

For example, having one developer on the most recent IDE version and another developer on a 2-month-old version can cause slight differences in debugging behavior that may lead to hours lost due to confusion. Similarly, a software library installed a year ago somewhere in the VS compiler and forgotten about can lead to compilation errors that offer unhelpful advice on the problem.

Local environments are suitable for interactive development, but the replicability issue should be top-of-mind. A common example of this type of environment: Every developer has a Windows machine with Keil MDK installed, using CMSIS as the base software, running tests on a development board connected to their PC. Code is shared through tools such as Git, but all development takes place locally.

Automated development such as CI tests is not suitable on local machines unless you are a solo developer working on your project alone. In team environments, the automated tests need to scale up, not depend on one individual machine being turned off, and be reproducible by any developer on the project. As the local environment's main problem lies with replicability, it should be ruled out for automated testing.

Virtual environments

Virtual environments create a dedicated environment for a specific purpose. The purpose could be general-purpose computing, such as having a **VirtualBox** that allows different OSes to run on one computer. In this context, virtual environments have the specific purpose to create an isolated environment with specific libraries and tools installed to avoid confusion. The isolated environment runs the *guest OS* that you develop on and is supported by the original *host OS* of your machine. For example, if you want to run a Windows program but you currently have a Mac machine, you can install VirtualBox on Mac, virtually run Windows as the guest OS, and run the program inside the Windows virtual environment.

Multipass is another, newer virtual environment specifically intended to create Ubuntu Linux environments quickly on your machine regardless of OS. **Docker** is a common virtual environment for developing small services for websites or servers and can be useful in the embedded context for developing or running code. Docker is unique in this list as it is a container, not a VM, meaning it shares the host OS's kernel and other components. The result of this difference is that Docker containers are smaller in size and start up much faster than VMs but are more restricted in certain functionality.

Here is a table breakdown summarizing the differences between these virtual environments:

	VirtualBox	**Multipass**	**Docker**
Host OS	Any	Any	Any
Guest OS	Any	Linux (Ubuntu)	Any
Image size	**Gigabytes (GB)**	GB	**Megabytes (MB)**
Startup time	Minutes	A minute	Seconds
Ideal for	Robust GUI development	Command-line general Linux development	Specific, repeatable services

Table 3.2 – Comparing virtual environments

These virtual environments can have some learning curve associated with them, though they are all well supported by documentation and a large ecosystem of users (with Multipass being the newest with the least support to date). The simplicity in usage comes after getting familiar with the environment and setting them up a few times.

Virtual environments are very good for code replicability as, typically, the environment is used specifically for one development purpose. All software, libraries, and tools installed in the virtual environment should serve that development purpose. To that end, longevity mainly comes from using the same type of environment with different software and tools installed for the project at hand.

Docker has the particular advantage of having an environment defined by code. A Dockerfile specifies the guest OS that will be the base of the container, and specific libraries and tools are installed line by line. This leads to excellent replicability of environments across teams and across times.

Interactive development typically requires a GUI, and thus VirtualBox is the go-to option in this group. Multipass and Docker can support applications with graphical interfaces but are generally designed for smaller tasks that are command-line-centric. A common example here would be a development team using an IDE that only works on Windows, but the team has a mix of Windows and Mac laptops. The Mac users can install a Windows VirtualBox environment and use the IDE inside that virtual environment along with the rest of their team.

Automated development such as CI compilations and tests are perfectly suited for virtual environments, particularly Docker. All dependencies are specified in code to reduce confusion, the startup times are short, and many Docker containers can be run in parallel on the same machine, leading to highly scalable regressions. In *Chapter 9, Implementing Continuous Integration*, we will discuss how to leverage Docker specifically in this CI context.

Cloud environments

Cloud environments refer to any environment you develop or that is running outside of your computer or laptop. This could be a network of servers your company owns, typically referred to as *on-prem*. Nowadays, it is getting more common to use cloud environments that are hosted by a different company. Amazon offers AWS, Google offers GCP, Microsoft offers Azure, Oracle offers **Oracle Cloud Infrastructure** (**OCI**), and Alibaba offers Aliyun. There are many services offered by each of these cloud providers and these can greatly enhance interactive and automated development when utilized effectively.

The primary downside of cloud environments today is simplicity. Learning how to use the various platforms can be time-consuming. While the different cloud provider services share some principles, switching cloud providers often requires relearning specific steps to be effective. This leads developers to often focus on one or two popular environments and use them for many future projects. AWS alone has dozens of services intended to simplify software development. Moreover, learning to use the correct option well takes time.

Once you learn how to work with cloud environments, they become powerful tools to expedite development. They can be configured to create highly replicable environments across teams or projects. Different services, once created, can be altered to serve new projects, leading to longevity in environments.

Cloud environments offer various services that streamline interactive and automated development. For example, you could use AWS EC2 instances to spin up virtual servers on demand for interactively coding, compiling, running on a virtual platform, and debugging, all through a graphical interface. These virtual servers have tools and libraries installed defined by code and can be saved as templates for further streamlining. You could use AWS CodeCommit to store your code base under development.

For automated testing, you could create Docker containers that automate regression compilations and tests and then store them in **AWS Elastic Container Service** (**AWS ECS**). They can then be automatically run when pushing code to your CodeCommit repository to validate code health.

Once tied into the AWS platform with these services, other offerings further enhance embedded product development and creation such as the AWS IoT Device Management service and AWS IoT Core software stack. This example is using a few specific services from AWS; there are more available from AWS and the other cloud providers that can greatly enhance embedded development.

One notable consideration in using cloud platforms is the potential difficulty of using physical hardware during development. If you are using a development board to run and debug code, it becomes more difficult to use cloud environments as they have no understanding of your local setup. Virtual platforms are ideal for developing code in cloud environments.

Summary

In this chapter, we covered the broad range of tools and software needed when developing Arm-based software. First, we analyzed software-centric and hardware-centric development platforms to execute code on. Next, we talked through the most common compiler toolchains and their components.

Thirdly, we explored different types of IDEs, some offered by Arm and others from around the Arm ecosystem. Lastly, we talked through different options for development environments—local, virtual, and cloud—in the context of interactive versus automated software development.

While there are many choices and choosing the right ones can seem overwhelming, this chapter should be a guide to selecting reasonable options for your project. If you end up not liking your pick, try something else the next time around. Newer ways of working are always being developed, especially innovation around cloud tools, so keep on the lookout for any emerging solutions.

This marks the end of *Part 1, Get Set Up*. You should now have a good understanding of the different hardware, software, and tools available to create a quality Cortex-M device. In the next part, each chapter will dive deep into specific topics such as optimization, security, and **machine learning** (**ML**). Every topic will be introduced in theory, with multiple examples provided per chapter to demonstrate concepts in practice.

Further reading

For more information, refer to the following resources:

- MPS3 prototyping board: `https://developer.arm.com/documentation/100765`
- VirtualBox documentation: `https://www.virtualbox.org/wiki/End-user_documentation`
- Multipass documentation: `https://multipass.run/docs`
- Docker usu cases overview: `https://www.docker.com/use-cases/`
- SystemC modeling language overview for the curious: `https://systemc.org/about/systemc/overview/`

Part 2:
Sharpen Your Skills

In this part, we will go through individual topics one by one, providing a short overview followed by multiple examples for you to learn by doing. You can skip chapters in this part based on your interest, but note that some examples build on the material in previous chapters. This part covers both software topics (booting up a device, optimizing performance, machine learning, and security) and software development topics (cloud development, continuous integration), with the final chapter containing helpful tips and future considerations.

This part of the book comprises the following chapters:

- *Chapter 4, Booting to Main*
- *Chapter 5, Optimizing Performance*
- *Chapter 6, Leveraging Machine Learning*
- *Chapter 7, Enforcing Security*
- *Chapter 8, Streamlining with the Cloud*
- *Chapter 9, Implementing Continuous Integration*
- *Chapter 10, Looking Ahead*

4
Booting to Main

The most prevalent language used in microcontroller development is C. The very first microcontrollers were 8-bit processors and programming was done in assembly language. Today, the microcontroller market is currently dominated by 32-bit processors, with Cortex-M holding a considerable market share. A significant amount of code is written in C, with some projects extending into C++. Python is popular for microcontroller applications and new languages such as Rust are also starting to appear, but a solid understanding of C programming is still a must for embedded developers. In this chapter, we will present how to get started with creating, building, running, and debugging C applications on Cortex-M microcontrollers. We will present the important parts of development with C. The focus is not on the language itself, but on the connection between the programming tools, the software, and the hardware.

The chapter is organized around the universal C program known as **hello world**, the smallest possible program to display a message. Since the introduction of the C programming language in the 1970s, every programming language has adopted the same approach to demonstrating an introductory program.

In embedded programming, there is a second universal program called **Blinky**. Blinky arises from the detail that many embedded systems powered by microcontrollers do not have hardware available to print messages. Instead of printing a message to a screen, Blinky lights an LED to demonstrate a small functional program.

We will demonstrate all the concepts covered using the three platforms introduced in *Chapter 1, Selecting the Right Hardware*:

- Arm Virtual Hardware using the Cortex-M55
- NXP LPC55S69-EVK using the Cortex-M33
- The Raspberry Pi Pico using the Cortex-M0+

The basics of hello world

Running hello world or Blinky on a microcontroller involves more than just a print statement. The traditional hello world program in C may look as follows:

```
#include <stdio.h>int main()
{
    printf("Hello Cortex-M world!\n");
}
```

If I compile this program and run it on any of my laptop computers, it will print the following message:

```
$ gcc -o hello hello.c
$ ./hello
Hello Cortex-M world!
```

Developers are not typically interested in the details of running a C program on a computer running Windows, Linux, or macOS. It's clear that the printf function is defined in a C header file, stdio.h, and there is a C library that provides the code. Most developers generally understand that a new process is created, the program is loaded into memory, and the C language defines the standard output file (stdio) to be the terminal and writes the message. The operating system will take care of the details to run the program and there is no real reason to think about how much or what kind of memory is used.

In embedded programming, the details taken care of by the operating system are more visible. In this chapter, we will present various ways of learning and understanding the details and how different programming environments will expose or hide these details. With a great SDK, you may never need to understand all of the details, but understanding the relationship between the hardware, the startup code used to initialize the hardware, the memory map, the ways to print messages, how to turn the LEDs on and off, and how to debug a program will enable you to create better products more quickly.

Let's go through each of the target platforms step by step and see what we can learn.

Arm Virtual Hardware using the Cortex-M55

To replicate the example in this section, you will be using the following tools and environment:

Platform	Arm Virtual Hardware – Corstone-300
Software	hello world
Environment	Amazon EC2 (AWS account required)
Host OS	Ubuntu Linux
Compiler	Arm Compiler for Embedded
IDE	-

Arm provides several reference designs for SoC designers to leverage when building their own Arm Cortex-M-based platforms. Corstone-300 is a reference design for a Cortex-M55 and Ethos-U55-based platform. In the previous chapter, we introduced virtual platforms and virtual IoT boards – what they are and how they are leveraged to accelerate software development for your target device without needing any physical hardware. Arm Virtual Hardware includes support for the Corstone-300 reference design. It can be accessed through an **Amazon Machine Image** (**AMI**) available on AWS Marketplace. Here is the link to get access: `https://aws.amazon.com/marketplace/pp/prodview-urbpq7yo5va7g`.

The AMI contains all the tools and licenses you will need to build and run software on the Corstone-300 virtual platform. The only prerequisite is an AWS account to launch an instance with this AMI.

Running hello world

Once you have started an instance with this AMI, you can clone this GitHub repository (`https://github.com/PacktPublishing/The-Insiders-Guide-to-Arm-Cortex-M-Development`) on your instance. This repository contains the software project files with instructions to build and run hello world on the Corstone-300 **Fixed Virtual Platform** (**FVP**) using Arm Compiler for Embedded. You can find the specific example for hello world on the Corstone-300 FVP under the `chapter-4` directory here: `https://github.com/PacktPublishing/The-Insiders-Guide-to-Arm-Cortex-M-Development/tree/main/chapter-4/hello-avh-corstone-300`.

The software project is Make-based. It contains the source files, linker file, and Makefile needed to build the hello world executable for this Cortex-M55-based system. To build `hello.axf`, which is our compiled executable, just run `make` from the command line:

```
git clone https://github.com/PacktPublishing/The-Insiders-
Guide-to-Arm-Cortex-M-Development/
cd The-Insiders-Guide-to-Arm-Cortex-M-Development/
chapter-4/hello-avh-corstone-300/make
make
```

With our software built, let us run `hello.axf` on the Corstone-300 FVP:

```
./run.sh -a hello.axf
```

You should see a few messages related to the FVP being launched followed by this:

```
Hello from Cortex-M55!
```

With that, you are using a virtual platform to run code on a Cortex-M55! While this is great, let us now dive deeper into the contents of this software project to get a better understanding of how the Arm Cortex-M55 processor is initialized, where code and data are placed in this system's memory map, and the mechanism that is used to print the **Hello from Cortex-M55!** message on the terminal.

Startup code

We start by inspecting the contents of the startup.c file. Regardless of your target platform, you need some startup code that is executed when the CPU resets. This file contains the initialization code – a list of things that need to be done at boot time by the processor and before entering the main application code. Let us break down this startup code:

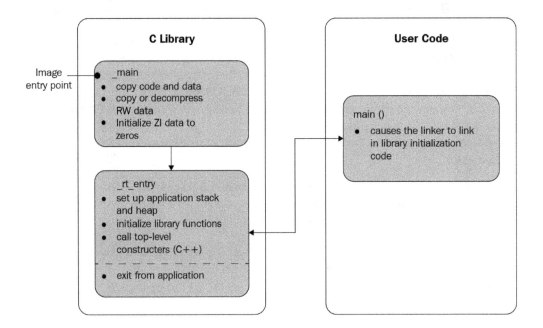

Figure 4.1 – An overview of the startup code

In the next section, we will break down and explain some of the essential elements of the startup code such as the exception vector table, reset handler, and C library main function. We will also dive into how certain hardware features such as an FPU and caches can be enabled in the startup code.

Exception vector tables and handlers

The initial stack pointer value and the exception handlers for all the different exception types that are supported by the Cortex-M55 processor are set up here. This includes the reset handler and exception handlers for things such as non-maskable interrupts and illegal or bad memory accesses, which are lumped under a category called `HardFault`.

The very first slice of data placed in flash memory is the exception vectors. They are placed in a special section of memory, called *BOOT* in this example. This named section will be used later by the scatter file to inform the linker that this section of data should be the very first one.

The reset handler and C library main function

As the name suggests, the reset handler is executed after the Cortex-M55 CPU comes out of reset. The reset handler calls the C library main function (__main):

```
__attribute__((interrupt)) void ResetHandler(void)
{
    __main(();
}
```

The C library main function (__main) should not be confused with the application main function (main). The C library __main function copies code and reads or writes data from the load region of your application to the execution region. These regions are described later, in the *Memory mapping* section of this chapter. The __main function also zeros out any uninitialized data in your application, creating consistency and eliminating erroneous data holding over from previous applications.

Enabling FPU and MVE support

If your software uses floating point operations and takes advantage of the MVE in the Cortex-M55 processor, you will need to explicitly enable hardware support for it in the startup code. The **Extension Processing Unit (EPU)** is the hardware bit on the Cortex-M55 CPU that controls both floating-point and MVE support.

To enable the EPU in both privileged and user modes of operation, we set the required bits of the **Coprocessor Access Control Register (CPACR)**:

```
#define CPACR (*((volatile unsigned int *)0xE000ED88))
CPACR = CPACR | (0xF << 20);
```

Enabling the branch cache

The **branch cache** is a hardware cache on the Cortex-M55 processor that is independent of the instruction and data caches. At reset time, the branch cache on the Cortex-M55 processor is disabled.

To maximize performance and take advantage of the Low Overhead Branch extension feature, we need to enable this cache by setting the appropriate bits in the CPU's **Configuration and Control Register (CCR)**. After writing the value to the CCR, an **instruction barrier (ISB)** must be executed to ensure that the effect takes place:

```
#define CCR_ADDR (0xE000ED14UL)
#define CCR *(volatile unsigned int *) CCR_ADDR
#define __ISB() __builtin_arm_isb(0xF)
CCR |= 0x00080000UL;
____ISB();
```

Initializing the caches

The Cortex-M55 processor includes support for optional instruction and data caches. In general, if you are running code or accessing data from an external memory connected to the main bus of the processor (which is very common), it is recommended to enable the caches. Both the instruction and data caches are disabled at reset. In our example, we enable the caches inside the main() function of our application in hello.c:

```
void cache_init()
{
    // Invalidate caches
    write_reg(ICIALLU, 0);

    // Enable both I & D caches
    write_reg(CCR, (read_reg(CCR) | 0x30000));
}
```

There are several other features on the Cortex-M55 processor that can be enabled at initialization time in your startup code. This includes things such as initializing the MPU, enabling the TCMs, and programming the **Security Attribution Unit (SAU)**. While this basic example does not demonstrate programming all the different registers on the CPU to enable this functionality, it should provide a good illustration of the bare basics required when writing your own startup code. You can find more information about these other features at the end of this chapter under the *Further reading* section.

We have looked at the important elements of the startup code needed to boot any Arm Cortex-M processor. Let us now walk through another essential aspect of building our embedded application – the placement of application code and data in the target device memory.

Memory mapping

If you do not define the memory map for your application, the linker in the compiler toolchain uses a default memory map for the placement of code and data sections. The syntax to provide this information to the linker depends on the toolchain that you use. In this section, we will use the Arm Compiler for Embedded toolchain to explain the important aspects of memory mapping. You can refer to the following link: https://developer.arm.com/documentation/100748/0618/Embedded-Software-Development/Default-memory-map.

This default memory map does not match that of our target, the Corstone-300 FVP. So, if you were to build this hello world application without specifying the memory map for the target, it would use the default one from the linker and would not run on the Corstone-300 FVP. To tell the linker what the memory map is for our target, we write a scatter file. The high-level memory map for our target is outlined here: https://developer.arm.com/documentation/101773/0001/Programmer-Model/System-Memory-Map-Overview/High-Level-System-Address-Map?lang=en.

Two types of memory regions are defined in a scatter file – load regions and execution regions. All your application code and data at reset time are contained in the load region. All other application code and data used while the application is running are placed in the execution region.

Let us inspect the different sections of the scatter file in our example and break down what each of these sections does, starting with the load region:

```
ROM0_LOAD 0x0 0x100000
{
    CODE +0
    {
        startup.o (BOOT, +First)
        * (+RO)
    }
}
```

ROM0_LOAD is the first load region that starts at the 0x0 address and has a maximum size of 0x100000. CODE is the first execution region. +0 after CODE denotes the offset value from the base address, 0x0. In this case, the offset is 0, so the region starts at 0x0. The exception vector table and handlers that we had marked as the *BOOT* section in our startup code are placed here in this execution region first, followed by all the other code and read-only data. Accesses in this memory region are performed either on the **Instruction Tightly Coupled Memory** (**ITCM**) or the CPU's main bus connected to external memory. The memory selected is dependent on whether the TCM interfaces are enabled on the Cortex-M55 CPU.

Next, let us look at what the RAM0_LOAD execution region specifies:

```
RAM0_LOAD 0x21000000 0x80000
{
    DATA +0
    {
        * (+RW,+ZI)
    }

    ARM_LIB_STACKHEAP 0x21070000  EMPTY 0x10000
    {}
}
```

RAM0_LOAD is the second execution region, which starts at the 0x21000000 address with a maximum size of 0x80000. The start address for this region, 0x21000000, is mapped to either the **Data Tightly Coupled Memory (DTCM)** or the first bank of SRAM memory available on our target. As with the previous region, which memory is used is dependent on whether the TCM interfaces are enabled on the Cortex-M55 CPU.

In this scatter file, we also define the stack and heap placement for our image at 0x21070000. This address also falls within the DTCM or SRAM region of our target. If you choose not to define the stack and heap placement in your scatter file as done here, you can instead re-implement the __user_setup_stackheap() function in your application.

Scatter files can get more complex than the one described here. There are several resources in the *Further reading* section if you would like to learn more about scatter files and the syntax for writing them, but this should give you a great start.

So far, we have looked at the initialization code and the scatter file, which informs the linker about the system's memory map. Let us now look at the I/O mechanism used in this example to print **Hello from Cortex-M55!** on the standard console.

I/O mechanisms

This example uses a mechanism called semihosting by default for **Input/Output (I/O)** functions. What is semihosting? semihosting enables software running on a target platform to directly access I/O facilities such as keyboard input or screen output on your host machine. For things such as console I/O and file I/O, you use the printf() and scanf() families of functions in the C library. The printf() and scanf() families of functions all eventually call the lower-level fputc() and fgetc() functions. The default implementation of fputc() and fgetc() uses semihosting.

While this can sound complex, `semihosting` provides a helpful mechanism to quickly debug and keep track of code through print statements with little setup or overhead. To understand how this process works in more detail, we can look at the *disassembly file* for the example software. A disassembly file essentially maps each location in memory to what instruction is stored there after compilation and interprets the instruction to be more human-readable. The disassembly file for this example is `hello.dis`.

`hello.dis` is created from the AC6 toolchain's `fromelf` utility. There's a rule in our Makefile that creates the disassembly file automatically by running `make`. To generate it standalone from the command line, run this:

```
fromelf -c hello.axf --cpu=Cortex-M55 > hello.dis
```

In `hello.dis`, look for any of the functions starting with *_sys*, such as `_sys_open`, `_sys_write`, `_sys_read`, or `_sys_close`. All these functions contain a BKPT instruction. BKPT disassembles as 0xbeXX, where XX is the breakpoint number. `semihosting` uses BKPT 0xab and hence, disassembles at 0xbeab. The type of `semihosting` operation can be determined by looking at the value passed in the `r0` register. The `r1` register contains all the other parameters that need to be passed in that operation. For example, in the following `_sys_write` disassembled block, we see MOVS   r0, #5, which means the operation is interpreted as a write and the output is sent to the system console:

```
_sys_write
        0x00000660:     b51f        ..      PUSH      {r0-r4,lr}
        0x00000662:     e88d0007    ....    STM       sp,{r0-r2}
        0x00000666:     4669        iF      MOV       r1,sp
        0x00000668:     2005        .       MOVS      r0,#5
        0x0000066a:     beab        ..      BKPT      #0xab
        0x0000066c:     b004        ..      ADD       sp,sp,#0x10
        0x0000066e:     bd10        ..      POP       {r4,pc}
```

All the `semihosting` operation types are documented here: https://github.com/ARM-software/abi-aa/releases/download/2022Q1/semihosting.pdf.

`semihosting` also comes in handy when you are working with a microcontroller that does not have any I/O functionality. The Corstone-300 FVP includes four **Universal Asynchronous Receiver-Transmitters (UARTs)** from the Arm **Cortex-M System Design Kit**. So, you can replace the C library's device driver-level functionality with your own implementation that is customized specifically for the UART on this system. This is referred to as *retargeting* the C library. In this example, we are going to use UART0 on the FVP for retargeting I/O.

To use this UART, first recompile your software in the make_uart directory by running the following:

```
cd The-Insiders-Guide-to-Arm-Cortex-M-Development/chapter-4/
hello-avh-corstone-300/make_uart
```

Now, run this newly compiled executable, hello.axf, on the Corstone-300 FVP, using the same command as before:

```
./run.sh
```

This time, the FVP will pop up a telnet terminal on UART0 with **Hello from Cortex-M55!**. Let us peek inside our code to understand what changes were made to get the UART working.

By defining the UART macro, the uart_config() function is called inside the application main() before any calls to printf():

```
#ifdef UART
        uart_config();
#endif
```

The uart_config() function is defined in serial.c, which contains the driver for the CMSDK UART. This function programs the transmit and receives bits on the UART's control registers to enable it and configures the baud rate setting:

```
void uart_config(uint32_t wUARTFrequency)
{
    CMSDK_UART0->CTRL = 0;              /* Disable UART when
changing configuration */
    CMSDK_UART0->BAUDDIV = wUARTFrequency / 115200ul;      /*
25MHz / 38400 = 651 */
    CMSDK_UART0->CTRL = CMSDK_UART_CTRL_TXEN_Msk|CMSDK_UART_
CTRL_RXEN_Msk;
}
```

The register addresses for the UART are defined in serial.h. The base address is set to 0x49303000, which maps to the address for UART0 on the Corstone-300 FVP's memory map:

```
#define CMSDK_UART0_BASE_ADDRESS              (0x49303000ul)
```

As `printf()` calls `fputc()`, we have redefined `fputc()` in `serial.c`, as shown in the following snippet, to use the UART. Ultimately, we have retargeted our application to use the UART:

```
int fputc (int c, FILE * stream)
{
    if (stream == &__stdout) {
        return (stdout_putchar(c));
    }

    if (stream == &__stderr) {
        return (stderr_putchar(c));
    }

    return (-1);
}
```

> **Important note**
> `sys_exit()` has been retargeted in `serial.c` to be an infinite loop. To end the simulation running on the virtual hardware, use *Ctrl + C* to terminate the process.

You now have a fundamental understanding of running a basic program on a virtual platform. The next examples will fill out your knowledge of startup code, memory mapping, and I/O mechanisms for different situations leveraging physical boards. Let us explore the Cortex-M33 through the NXP board next.

NXP LPC55S69-EVK using the Cortex-M33

To replicate the example in this section, you will be using the following tools and environments:

Platform	NXP LPC55S69-EVK
Software	Blinky
Environment	Personal Computer
Host OS	Windows
Compiler	Arm Compiler for Embedded
IDE	Keil MDK Community

The **LPCXpresso55S69 EVK** is a development board from NXP with an MCU based on the Arm Cortex-M33. The board comes with a debug probe, audio subsystem, and accelerometer. You can also add several off-the-shelf add-on boards for sensors and networking. Here's the link with all the specifications for this board: `https://www.nxp.com/design/development-boards/lpcxpresso-boards/lpcxpresso55s69-development-board:LPC55S69-EVK`.

As we talked about in the previous chapter, once you have selected your target board, you are presented with several different SDK options. The software project you start with is often dependent on the SDK you choose. This board is supported by the MCUXpresso SDK from NXP, Keil MDK from Arm, and IAR Workbench from IAR Systems, to name a few. The functionality and features supported vary by SDK. In this example, we are going to use **MDK-Community**, a free version of Keil MDK for hobbyists to run the Blinky-led application on this board.

In MDK, support for several boards and software components is provided via CMSIS-Pack. CMSIS-Pack is a distribution mechanism for software components, devices, and board support files. It includes source code, header files, startup code, software libraries, and ready-to-use software examples that can be deployed onto your target board. These examples range from the simple hello world and Blinky software to more complex applications that demonstrate securely connecting the board to the cloud, sending messages, and checking the device status in the cloud. If your board has a CMSIS-Pack available as this NXP board does, it is more than likely that you will use that instead of starting your software project from scratch.

Running Blinky

In this section, we will focus on the simple `led_blinky` example provided through the CMSIS-Pack for NXP LPC55S69-EVK. There are detailed instructions to get the CMSIS-Pack for this board and run the example software projects on it with Keil MDK tools. Go to this link and follow the detailed instructions through running the `led_blinky` application on your board (pages 3 to 11): `https://developer.arm.com/documentation/kan322/latest/`.

In summary, you should be performing the following high-level steps:

1. First, download a copy of MDK-Community on your PC. Note that MDK is only supported on Windows. You will need to get an Arm account from `developer.arm.com`. Once you log in with your Arm account, go to `https://www.keil.arm.com/mdk-community/`, download MDK-Community, and follow the instructions given to get a free license key that you can use for a year.

2. Next, download the Pack for this board from `https://www.keil.com/boards2/nxp/lpcxpresso55s69` and install it using the Pack installer included with MDK. Then, copy the `led_blinky` example into a μVision project. Alternatively, you can also use the Pack Installer to directly locate the example and copy it into your workspace.

3. Power the board by connecting the USB cable provided with your board to the P6 Debug Link connector.

4.   Finally, build the led_blinky application in µVision and load it into the LPC55S69 Flash memory and run it.

After following these steps (the Keil link containing very detailed instructions with pictures), you should have the LED successfully blinking on the board.

Next, let us dive into the software contents of this CMSIS-Pack and inspect the startup code, linker files, and I/O mechanisms used in this example.

Startup code

With a CMSIS-Pack, the startup code comes from the CMSIS-CORE software component and is typically structured as shown here:

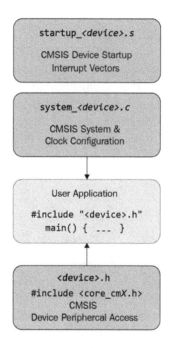

Figure 4.2 – The structure of CMSIS-CORE

The startup_LPC55S69_cm33_core0.S file is an assembly file that contains all the exception and interrupt vectors, implementing a default function for every interrupt on this board. The stack and heap for the application are also configured in this file. This startup code is executed after reset and calls the SystemInit() function, which is defined in the system_LPC55S69_cm33_core0.c file. It enables the FPU and compiler support for the Security Extensions on the Cortex-M33 processor. Details on the Security Extensions are provided in the *Further reading* section.

The processor clock speeds are set up in this system-specific initialization file. The `led_blinky` example uses the **System Timer (SysTick)** for the delay function so that the LED can blink every 1,000 ms. For this to work, SysTick on the Cortex-M33 processor needs to be configured to start a periodic timer interrupt. The configuration of SysTick is done in the `LPC55S69_cm33_core0.h` file. This header file also provides a standardized register layout for all the peripherals on this board.

Memory mapping

A major advantage of using CMSIS-Pack is the simplification of allocating instruction and data memory manually. Each board supported by a CMSIS-Pack contains the required memory mapping files, supporting multiple compilers, to simplify the process of getting started. If you would like to reference the scatter file or linker file detailing the memory map for this example, you can find it under the appropriate toolchain folder of the imported project.

I/O mechanisms

The `led_blinky` software example does not use or need an I/O mechanism, as all it does is provide a heartbeat for the board and blink an LED. It does not have any `printf` statements in it. You can, however, easily modify your application to add a `printf` statement and use the **Debug Viewer** window on µVision to view the output. To retarget the output in this example, we use a utility called Event Recorder that µVision provides. We need to include the `EventRecorder` header file in our code first:

```
#include "EventRecorder.h"
```

We then need to add these two lines of code to initialize and start Event Recorder in the `main()` function:

```
EventRecorderInitialize (EventRecordAll, 1);
EventRecorderStart ();
```

For more detailed instructions on how to implement this feature in Keil MDK, refer to the event recorder at `https://www2.keil.com/mdk5/debug/eventrecorder`. As with `semihosting`, this I/O mechanism does not require hardware such as a UART and can be implemented with few code changes. `EventRecorder` provides an API that you can use to record different events in your code in a file on your target platform.

Another commonly used I/O mechanism to output data is via the UART. This mechanism is used by several example applications included for this board, such as `hello_world`. First, ensure your board is connected via the USB cable to the Debug Link port (P6) on the board. Then identify the COM port that your board is using for the serial communication. To do this, open **Device Manager** on your Windows PC and expand the **Ports (COM & LPT)** section, as shown here:

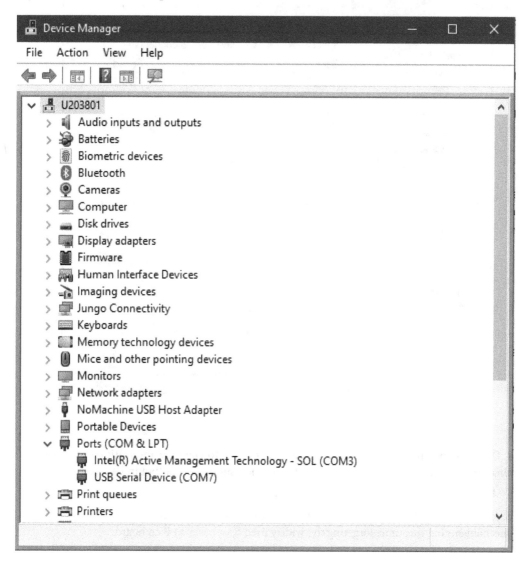

Figure 4.3 – Device Manager

In our case, it is using **COM7**. Now, using a terminal application (we like PuTTY), open a serial connection on **COM7** with a **Speed** setting of 115200:

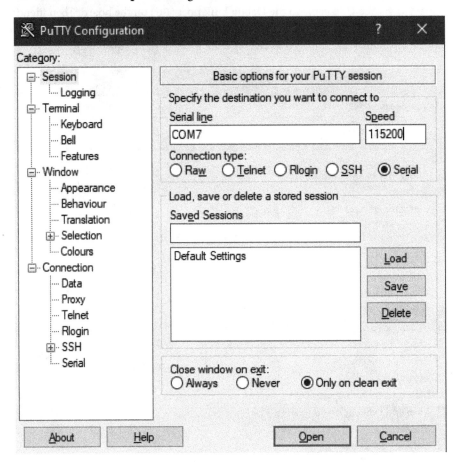

Figure 4.4 – Opening a serial connection

Click **Open** to open the serial connection. You will be able to view the application output on the terminal window that opens up. We will use this mechanism to view output in later chapters, so it's best to get familiar with it.

In the final example, let us look at the same primary elements – startup code, memory mapping, and I/O mechanisms but this time, targeting the widely used Raspberry Pi Pico board.

The Raspberry Pi Pico using the Cortex-M0+

To replicate the example in this section, you will be using the following:

Platform	The Raspberry Pi Pico
Software	hello world
Environment	Raspberry Pi 4
Host OS	Linux (Ubuntu)
Compiler	GCC
IDE	OpenOCD

The foundation of software development in C on the Raspberry Pi Pico is the Pico SDK. It provides header files and libraries to enable simple access to all the Pico hardware. Running a C program on the Pico is both similar to and different from running hello world on Windows, Linux, or macOS. The main similarity is that the C programs created with the Pico SDK also start with a main function. The core difference is that the Pico only runs one program at a time. There is no operating system to start, schedule, and stop programs. There is just a single program running on Cortex-M processors at one time.

Now, let's dive in and run hello world on the Raspberry Pi Pico.

Running hello world and Blinky

The first step towards getting hello world running is to set up the Pico SDK. The SDK uses the GCC compiler and cbuild to create applications. The *Getting started with Raspberry Pi Pico C/C++ development* manual is the best place to start and is very detailed. You can find it here:

https://datasheets.raspberrypi.com/pico/getting-started-with-pico.pdf

A straightforward way to work with the Raspberry Pi Pico is through the Raspberry Pi 4 (or Raspberry Pi 400). There is a single script that can be run to install the necessary development tools and set up the SDK. The script is called pico_setup.sh. The script installs some software using apt install, clones the SDK repositories from GitHub, builds a few examples, and installs additional tools, including VS Code and openocd for debugging.

Here are the commands you will need to get and install the pico_setup.sh script:

```
$ wget https://raw.githubusercontent.com/raspberrypi/pico-
setup/master/pico_setup.sh
$ chmod +x pico_setup.sh
$ ./pico_setup.sh
```

There are also other ways to get started, such as from a PC on Linux, Windows, or Mac. If you do not have a Raspberry Pi 4 or prefer to code from your PC, refer to the latest *Getting started* guide (https://datasheets.raspberrypi.com/pico/getting-started-with-pico.pdf). The Raspberry Pi 4 approach we use makes debugging using a UART easier in the following sections.

Before proceeding with hello world, make sure the required tools are operational and check the installed versions:

- Compiler:

```
$ arm-none-eabi-gcc --version
arm-none-eabi-gcc (15:9-2019-q4-0ubuntu1) 9.2.1 20191025
(release) [ARM/arm-9-branch revision 277599]
```

- Debugger:

```
$ gdb-multiarch --version
GNU gdb (Ubuntu 9.2-0ubuntu1~20.04.1) 9.2
Copyright © 2020 Free Software Foundation, Inc.
```

- cmake:

```
$ cmake --version
cmake version 3.16.3

CMake suite maintained and supported by Kitware (kitware.
com/cmake).
```

- openocd:

```
$ openocd --version
Open On-Chip Debugger 0.11.0-g610f137-dirty (2022-03-26-
14:09)
Licensed under GNU GPL v2
For bug reports, read
        http://openocd.org/doc/doxygen/bugs.html
```

Now that you are more familiar with the Cortex-M boot code, we can get into further detail by creating a hello world example from scratch. There are similar examples available, but building from scratch is a great way to reinforce your understanding of the fundamentals to create better code in the future.

Starting from scratch

In this section, we will start simply with a blinking LED and print statements over a USB. After that, we will print the output to the UART. Finally, we will connect the debugger and have a closer look at the memory map, hardware setup code, and then a multicore example.

Here is the source code for `hello-world.c`:

```c
#include <stdio.h>
#include "pico/stdlib.h"

int main()
{
    const uint LED_PIN = PICO_DEFAULT_LED_PIN;

    gpio_init(LED_PIN);
    gpio_set_dir(LED_PIN, GPIO_OUT);

    stdio_init_all();

    while (true)
    {
        printf("Hello, Pi Pico!\n");
        gpio_put(LED_PIN, 1);
        sleep_ms(500);
        gpio_put(LED_PIN, 0);
        sleep_ms(500);
    }

    return 0;
}
```

Copy the preceding code into a text file named `hello.c` using a text editor. The project is also available on GitHub under the following directory: `https://github.com/PacktPublishing/The-Insiders-Guide-to-Arm-Cortex-M-Development/tree/main/chapter-4/hello-pico`.

Besides the source file, a CmakeLists.txt file is needed to specify how to build the source code with cbuild. Copy the following text and save it as CmakeLists.txt:

```
cmake_minimum_required(VERSION 3.18)

include($ENV{PICO_SDK_PATH}/external/pico_sdk_import.cmake)

project(hello C CXX ASM)
set(CMAKE_C_STANDARD 11)

pico_sdk_init()

add_executable(${PROJECT_NAME}
            ${PROJECT_NAME}.c
        )

# pull in common dependencies
target_link_libraries(${PROJECT_NAME} pico_stdlib)

# enable usb or uart output
pico_enable_stdio_usb(${PROJECT_NAME} 1)
pico_enable_stdio_uart(${PROJECT_NAME} 0)

# create map/bin/hex/uf2 file etc.
pico_add_extra_outputs(${PROJECT_NAME})
```

Next, use an environment variable to find the SDK. Set the path to your SDK using the environment variable as follows. This line can also be added to your ~/.bashrc file so it's always set:

```
$ export PICO_SDK_PATH=/home/pi/pico/pico-sdk
```

We are now ready to build the hello world program. It is best to run cmake from a new directory. This makes it easy to delete the build directory anytime to get a fresh start. You can do this using the following commands:

```
$ mkdir build ; cd build
$ cmake ..
$ make
```

The quickest way to run the program is to copy the resulting `hello.uf2` file to the Pico, and it will load and run.

To run the program, unplug the USB cable, hold down the **BOOTSEL** button, and plug in the USB cable. This will cause the Pico to appear as a USB storage device on your computer. The path will be different depending on your operating system. The path used in the following `copy` command is for Raspberry Pi OS:

```
$ cp hello.uf2 /media/pi/RPI-RP2/
```

When the file is copied, the Pico will disappear as a storage device from your computer and the hello world program will start running. You should see the LED on the Pico blinking as specified in `hello.c`.

Next, let's find the output from the `printf` statement in `hello.c`.

The print statement with the `hello` string is directed to the USB serial. This is because the `CMakeLists.txt` file sets the USB serial as the default with this line:

```
pico_enable_stdio_usb(${PROJECT_NAME} 1)
```

To connect to the USB serial, use `minicom`:

```
$ minicom -b 115200 -o -D /dev/ttyACM0
```

The terminal will then start to output the `hello` string:

```
Welcome to minicom 2.8

OPTIONS: I18n
Port /dev/ttyACM0, 14:24:31

Press CTRL-A Z for help on special keys

Hello, Pi Pico!
Hello, Pi Pico!
Hello, Pi Pico!
Hello, Pi Pico!
```

To exit `minicom`, use *Ctrl + A* and then *X* and hit return for *yes*.

You now have run both the hello world and Blinky examples coded from scratch on a Cortex-M board! The next sections add further detail to the concepts of memory mapping and I/O that we already introduced.

I/O mechanisms

Using the UART is more complex, as it requires physical pins to be connected. The easiest way is to use a Raspberry Pi 4 (which is why we chose this path to start with).

Connect the **ground** (**GND**) and 2 GPIO pins of the Raspberry Pi 4 to the GND and 2 GP pins on the Pico. There is a diagram of how to do this in the *Getting Started* guide referenced earlier (https://datasheets.raspberrypi.com/pico/getting-started-with-pico.pdf). Once the pins are connected, you can make the changes to the build, rebuild the software, and run it again with the UART.

To change from the USB serial to the UART, change CMakeLists.txt to enable uart (1) and disable usb (0):

```
pico_enable_stdio_usb(${PROJECT_NAME} 0)
pico_enable_stdio_uart(${PROJECT_NAME} 1)
```

After making the change, delete the build directory and run cbuild and make again:

```
$ rm -rf build
$ mkdir build ; cd build
$ cmake ..
$ make
```

Use the same procedure to unplug the USB cable from the Pico: hold down the **BOOTSEL** button, plug in the USB cable, and copy the hello.uf2 file to the storage device. The LED on the Pico will start blinking again.

To see the UART, run minicom again, but this time, with the serial0 device:

```
$ minicom -b 115200 -o -D /dev/serial0
```

The terminal will start to output the hello string:

```
Welcome to minicom 2.8

OPTIONS: I18n
Port /dev/serial0, 15:30:49

Press CTRL-A Z for help on special keys

Hello, Pi Pico!
Hello, Pi Pico!
```

```
Hello, Pi Pico!
Hello, Pi Pico!
```

So far, we have mastered the basics of the LED, USB serial, and UART. You are probably thinking it's a hassle to unplug the Pico, hold down the **BOOTSEL** button, and plug it in again every time you want to change the software, especially if you are not sitting next to the board. We think so too and the next section covers how to make this process more streamlined.

Loading the program using the debugger

Instead of loading using the **BOOTSEL** button and copying the file, it's possible to load the software using a debugger connection. Connecting a debugger requires connecting the 3 SWD pins on the Pico to either the GPIO of a Raspberry Pi 4 or to another debug probe. I will explain how to do it with a Raspberry Pi 4, but it can also be done with a second Raspberry Pi Pico or even on the same Pico because there is a second Cortex-M0+ that can be used to run the debug probe software:

Raspberry Pi 4	Raspberry Pi Pico
GND (Pin 20)	SWD GND
GPIO24 (Pin 18)	SWDIO
GPIO25 (Pin 22)	SWCLK

Table 4.1 – Mapping the GPIO pins on the Raspberry Pi 4 to the SWD pins on the Raspberry Pi Pico

Once the hardware connections are made, restart the board and we can load applications from the command line with no need to power the Pico and hold down the **BOOTSEL** button anymore. The openocd program will load the configuration files for the board and run a sequence of commands in the double quotes. The following example shows how to program the board with hello.elf, verify the programming, reset the board, and then exit. Notice that the .elf file is used instead of the .uf2 file:

```
$ openocd -f $PICO_SDK_PATH/../openocd/tcl/interface/
raspberrypi-swd.cfg -f $PICO_SDK_PATH/../openocd/tcl/target/
rp2040.cfg -c "program build/hello.elf verify reset exit"
```

Interactive debugging can also be done using gdb. In one window, start openocd. It's the same as before, but with no commands:

```
$ openocd -f $PICO_SDK_PATH/../openocd/tcl/interface/
raspberrypi-swd.cfg -f $PICO_SDK_PATH/../openocd/tcl/t
arget/rp2040.cfg
```

In another window, use gdb to debug:

```
$ gdb-multiarch build/hello.elf
(gdb) target remote localhost:3333
Remote debugging using localhost:3333
warning: multi-threaded target stopped without sending a
thread-id, using first non-exited thread
0x10000b2e in sleep_until ()
```

Download the hello.elf image to debug:

```
(gdb) load
Loading section .boot2, size 0x100 lma 0x10000000
Loading section .text, size 0x2310 lma 0x10000100
Loading section .rodata, size 0xf8 lma 0x10002410
Loading section .binary_info, size 0x24 lma 0x10002508
Loading section .data, size 0x180 lma 0x1000252c
Start address 0x100001e8, load size 9900
Transfer rate: 18 KB/sec, 1980 bytes/write.
```

Before running any other command, perform a reset:

```
(gdb) monitor reset init
target halted due to debug-request, current mode: Thread
xPSR: 0xf1000000 pc: 0x000000ee msp: 0x20041f00
target halted due to debug-request, current mode: Thread
xPSR: 0xf1000000 pc: 0x000000ee msp: 0x20041f00
```

Set a breakpoint on the main function:

```
(gdb) b main
Breakpoint 1 at 0x10000366
```

Now, we can load a program using gdb, set breakpoints, single step through source code, and examine registers and memory.

Having gone over how to load and debug the application on the Raspberry Pi Pico and make the required hardware connections, let us now inspect the startup code and the code used to access the UARTs for I/O.

Startup code

As with the other examples, we want to identify the very beginning of the software. In this case, that can be found in the SDK file at `src/rp2_common/pico_standard_link/crt0.S`.

Here, we have the `_reset_handler` function. It includes a call to `runtime_init()` and then `main()`. The `runtime_init()` function initializes all the hardware peripherals and prepares them for use.

The memory map is a fairly standard Cortex-M memory map. The following table lists the start address for the different regions on the Raspberry Pi Pico board:

Memory Region	Start Address
ROM	0x00000000
XIP (Flash execute in place)	0x10000000
SRAM	0x20000000
Peripherals connected to **Advanced Peripheral Bus (APB)**	0x40000000
Peripherals connected to APB	0x50000000
IOPORT registers	0xd0000000
Cortex-M0+ internal registers	0xe0000000

Table 4.2 – The memory map for the Raspberry Pi Pico

The linker files for the `gcc` compiler are in `src/rp2_common/pico_standard_link`.

There are four different `.ld` files:

- `Memmap_default.ld`: Default value to put code in Flash
- `memmap_no_flash.ld`: Puts the code straight into SRAM
- `memmap_blocked_ram.ld`: Puts the code in striped banks of RAM
- `memmap_copy_to_ram.ld`: Copies the code from Flash to RAM

To find out more details about the compilation, use the verbose output. After `cmake`, run `make` using the following:

```
$ make VERBOSE=1
```

This will generate all the details of the compiler flags, linker files, and much more. Through these, you can learn what is happening during the compilation.

I/O mechanisms (using the UARTs this time)

The Pico has 2 Arm PL011 UARTs. The SDK provides an easy way to access UARTs.

Look in `src/rp2_common/hardware_uart` for the `uart.c` and `uart.h` files to see the details on how to access the UARTs.

Here is an example program to print to the UART directly:

```
#include "pico/stdlib.h"

int main()
{

    // Initialise UART 0
    uart_init(uart0, 115200);

    // Set the GPIO pin mux to the UART - 0 is TX, 1 is RX
    gpio_set_function(0, GPIO_FUNC_UART);
    gpio_set_function(1, GPIO_FUNC_UART);
    uart_puts(uart0, "hello world!\r\n");

    return 0;
}
```

Debugging tips

For each software application, remember to look at the `.dis` file in the `build` directory. This is the disassembly file you can use to look at the instructions. To make this easier to understand, build the application with the debug information.

When running `cmake`, add `CMAKE_BUILD_TYPE`:

```
$ cmake -DCMAKE_BUILD_TYPE=Debug ..
```

Now, run `make` as normal:

```
$ make
```

You can use `objdump` to get a file that mixes the source code and assembly instructions. This makes it far easier to understand which lines of C code map to which instructions:

```
$ arm-none-eabi-objdump -S hello.elf > hello.mix
```

Now, open `hello.mix` in a text editor, and search for `main`. Find a nice display of C and assembly:

```
int main()
{
1000035c:       b570            push    {r4, r5, r6, lr}
    const uint LED_PIN = PICO_DEFAULT_LED_PIN;

    gpio_init(LED_PIN);
1000035e:       2019            movs    r0, #25
10000360:       f000 f834       bl      100003cc <gpio_init>
 * Switch all GPIOs in "mask" to output
 *
 * \param mask Bitmask of GPIO to set to output, as bits 0-29
 */
static inline void gpio_set_dir_out_masked(uint32_t mask) {
    sio_hw->gpio_oe_set = mask;
10000364:       23d0            movs    r3, #208        ; 0xd0
10000366:       061b            lsls    r3, r3, #24
10000368:       2280            movs    r2, #128        ; 0x80
1000036a:       0492            lsls    r2, r2, #18
1000036c:       625a            str     r2, [r3, #36]   ; 0x24
```

As anyone developing software knows, getting it right on the first try is unlikely. These debugging tips should help when porting your own application to the Pico board.

Summary

This chapter outlined the basics of getting started with developing code for Cortex-M devices in practical terms. We got basic software running on a variety of Cortex-M-based platforms on a variety of environments and IDEs. Each example examined startup code, memory mapping, and I/O mechanisms in progressively more detail while offering tips for effectively debugging code throughout.

The skills learned in this chapter will help you start your next Cortex-M project from a position of fundamental knowledge, eliminating the black magic of initializing hardware. You should now be aware of common setup issues and be able to iterate to a solution more quickly, helping you get to the fun parts of developing that much quicker.

The fun part of coding, to us at least, is creating code that interacts with the real world in exciting ways. Machine learning applications offer incredible opportunities for impacting the world through smaller devices and are moving away from the cutting edge to become more mainstream. The next chapter investigates how to use machine learning in your next Cortex-M project, from setting up to running inferences.

Further reading

- The Cortex-M55 Technical Reference Manual, which adds more context for features such as the CCR: `https://developer.arm.com/documentation/101051/0002`.

- A discussion of the `__main` function, found in the Arm Compiler for Embedded documentation: `https://developer.arm.com/documentation/100748/0618/Embedded-Software-Development/Application-startup`.

- An explanation of a low overhead loop can be found in the Armv8.1-M architecture reference manual: `https://developer.arm.com/documentation/ddi0553/bs/`.

- A basic tutorial explaining `cbuild` and `cmake` procedures: `https://cmake.org/cmake/help/latest/guide/tutorial/index.html`.

- An introduction to Security Extensions for the Cortex-M33: `https://developer.arm.com/documentation/100235/0100/Introduction/About-the-Cortex-M33-processor-and-core-peripherals/Security-Extension?lang=en`

- Register reference for the Cortex-M55, helpful for addressing necessary registers and components at startup: `https://developer.arm.com/documentation/101273/r0p2/Cortex-M55-Processor-level-components-and-system-registers---Reference-Material`.

- A high-level overview of scatter file syntax to work with Arm Compiler for Embedded startup code: `https://developer.arm.com/documentation/dui0474/m/scatter-file-syntax`.

- How to get the Arm Compiler for Embedded linker, `armlink`, and the scatter file syntax to work together effectively: `https://developer.arm.com/documentation/dui0474/m/scatter-loading-features`.

5
Optimizing Performance

Optimizing software for performance is an important topic for Cortex-M software development. Obtaining ample application performance from a particular Cortex-M device is essential to avoid selecting a processor with a higher cost, larger silicon area, or more power than needed. Optimizing performance can be used to select the correct processor for a given use case and can also be used after selecting a processor to optimize system performance in general.

In *Chapter 1*, *Selecting the Right Hardware*, we looked at the wide range of Cortex-M processors and reviewed use cases and hardware characteristics to help determine the right processor for an application. Processor selection, however, is only one variable contributing to building a system with optimal performance. There are two other key factors: compiler settings and algorithm implementation. Measuring performance and making informed changes to these three variables will lead to a solid design with excellent performance.

In this chapter, we use one example software use case (a simple dot product calculation of two vectors) and analyze how its performance is affected by changing the processor, algorithm implementation, and compiler options. In the process, we will demonstrate a method for taking performance measurements of a critical section of code.

In a nutshell, the following topics will be covered:

- Our algorithm – the dot product
- Measuring cycle count
- Measuring dot product performance
- Optimization takeaways

Our algorithm – the dot product

We will use the dot product, also called the scalar product, as a straightforward algorithm to clearly demonstrate the concepts of performance optimization. The dot product of vectors provides information about the lengths and angles of vectors and is frequently used in ML applications. The dot product is a very easy calculation for teaching purposes. It also can be done in multiple ways and can take advantage of vector processing hardware. In fact, the Arm Cortex-A processors have special instructions to increase the performance of dot product computation (the Cortex-M processors, as of yet, do not).

Of course, in a realistic setting, your software algorithms will be more complex than a simple dot product. The same underlying principles of optimization will apply, however, as there are many ways to solve any given problem in software. Understanding how to compare these implementations quickly will enable effective system optimization.

As a quick review of what the dot product does, let's take two vectors and calculate the dot product:

```
V1 = [1, 3, -5]
V2 = [4, -2, -1]
```

First, we multiply the vectors to create a new vector:

```
V3 = [4, -6, 5]
```

Finally, we sum the values of the vector to produce a scalar value:

```
DP = 4 + (-6) + 5
DP = 3
```

The following is a simple C program using integers to calculate the dot product for vectors v1 and v2. Copy the text into a file called dot-simple.c or get it from the book's GitHub project: (https://github.com/PacktPublishing/The-Insiders-Guide-to-Arm-Cortex-M-Development/tree/main/chapter-5/dotprod-personal-computer):

```c
#include <stdio.h>
int dot_product(int v1[], int v2[], int length)
{
    int sum = 0;
    for (int i = 0; i < length; i++)
        sum += v1[i] * v2[i];
    return sum;
}


int main(void)
{
    int len = 3;
```

```
    int v1[] = {1, 3, -5};
    int v2[] = {4, -2, -1};

    int dp = dot_product(v1, v2, len);
    printf("Dot product is %d\n", dp);
    return 0;
}
```

On a computer with a C compiler, build and run it. The following commands demonstrate using GCC to compile and run:

```
$ gcc dot-simple.c -o dot-simple
$ ./dot-simple
Dot product is 3
```

The authors' recommended quick and free C compiler options are listed here, by OS, for your convenience:

- Windows: MSYS2 (`https://www.msys2.org/`)
- Linux: GCC (installed by default on most distributions)
- Mac: Clang via Xcode

With this understanding of the dot product example, let's see how to measure the cycles required to compute the dot product on a Cortex-M microcontroller.

Measuring cycle count

Cortex-M processors provide multiple ways to compare the performance of different implementations of an algorithm. One way is to use a cycle counter register. Another way is to use a timer to measure the time of execution and convert the time into clock cycles using the processor frequency. The most recent Cortex-M microcontrollers contain a full **performance monitoring unit (PMU)**, which enables software to get information about the count of various events occurring while the software is executing. One of the measured events can be a **cycle counter**, but additional events such as cache and memory accesses can also be counted.

This section will cover how to use the most common methods of counting cycles for a specific section of code on Cortex-M microcontrollers. First, we will introduce **System Tick Timer** or **System Time Tick (SysTick)** followed by **Data Watchpoint and Trace (DWT)**. Both these programming interfaces can be used to count clock cycles for a benchmark or section of software. Examples are given in context later in this chapter.

System Tick Timer

SysTick is a system timer peripheral, included inside a Cortex-M processor. It includes a count-down timer and can generate interrupts to the processor core itself. It is used for operating system context switching or timekeeping. It can also be used to measure time to find out how long a section of code takes to execute.

SysTick has a simple programming interface that is used to set up the timer. The three primary registers used to control the timer are the following:

- **Control and status register**: Used to configure, start, and stop the timer
- **Reload value register**: Used to load a value into the counter
- **Current value register**: Returns the current value of the counter

The counter is 24 bits and can measure any software that runs for less than 16,777,215 clock cycles. As an example, for hardware running at 100 MHz, SysTick can measure software running up to 168 milliseconds without overflowing the counter.

To create a simple interface to SysTick, use the `systick.h` file from the book's GitHub project: `https://github.com/PacktPublishing/The-Insiders-Guide-to-Arm-Cortex-M-Development/tree/main/chapter-5/dotprod-pico/systick.h`. The contents of the file are shown here for convenience:

```c
#include <stdint.h>

void start_systick(void);
void stop_systick(void);

/* Systick variables */
#define SysTick_BASE            (0xE000E000UL +   0x0010UL)
#define SysTick_START           0xFFFFFF

#define SysTick_CSR             (*((volatile uint32_t*)(SysTick_
BASE + 0x0UL)))
#define SysTick_RVR             (*((volatile uint32_t*)(SysTick_
BASE + 0x4UL)))
#define SysTick_CVR             (*((volatile uint32_t*)(SysTick_
BASE + 0x8UL)))

#define SysTick_Enable          0x1
#define SysTick_ClockSource     0x4
```

The systick.h file contains the location of the peripheral registers in the Cortex-M memory map, the constants to use for programming, as well as the function prototypes to start and stop the counter.

The control logic is found in the systick.c file, also located in the book's GitHub project. The contents of the file are shown here for convenience:

```
#include <stdio.h>
#include "systick.h"

void start_systick()
{
    SysTick_RVR = SysTick_START;
    SysTick_CVR = 0;
    SysTick_CSR |=  (SysTick_Enable | SysTick_ClockSource);
}

void stop_systick()
{
    SysTick_CSR &= ~SysTick_Enable;

    uint32_t cycles = (SysTick_START - SysTick_CVR);

    printf("CCNT = %u\n", cycles);
    if (SysTick_CSR & 0x10000)
        printf("WARNING: counter has overflowed, more than
16,777,215 cycles");

}
```

This file contains a function to start the timer and another to stop the timer and read the cycle count. The start function sets the clock source as the processor clock.

You can now use SysTick in a Cortex-M application to count clock cycles by wrapping a section of code with the start and stop functions. The cycles it takes will automatically print from the printf statement included in the stop_systick function. To test this, try putting the start and stop functions around printf() to measure how many cycles it takes:

```
#include <stdio.h>
#include "systick.h"
```

```
int main()
{
    (void) start_systick();

    printf("Count this hello world using SysTick\n");

    (void) stop_systick();
}
```

Note that the preceding code will only run on a Cortex-M microcontroller with SysTick so don't try to run it on your laptop! We will use this SysTick measurement method later in this chapter on Cortex-M devices.

Next, let's look at DWT as another way to count clock cycles.

Data Watchpoint and Trace

Another way to count clock cycles on Cortex-M microcontrollers is using DWT. Not all Cortex-M microcontrollers have DWT implemented, but most do. The Cortex-M0+ used in the Raspberry Pi Pico does not include DWT for cycle count measurement.

DWT provides functionality beyond counters. It includes hardware watchpoints and triggers for debugging. Our focus will be on using the cycle counter and the additional counters that report various hardware events. If interested, refer to the technical reference manual for a particular Cortex-M processor to learn about the full DWT features and complete set of registers.

DWT includes a list of six counters that are easy for software to access when measuring performance. Here is the list:

- Clock cycle count
- Number of folded instructions
- Cycles performing loads and stores
- Cycles sleeping
- Count of instruction cycles beyond the first cycle (CPI cycles)
- Cycle spent processing interrupts

Using DWT requires a special data write to the **lock access register** (**LAR**). This is the most common mistake programmers make when trying to use DWT functionality. To create a simple interface to DWT, refer to the dwt.h file in the book's GitHub project: https://github.com/PacktPublishing/The-Insiders-Guide-to-Arm-Cortex-M-Development/tree/main/chapter-5/

dotprod-nxp-lpcxpresso55s69/dwt.h. As with the SysTick files, the dwt.h file contents are displayed here for convenience:

```
#include <stdint.h>


void start_dwt(void);
void stop_dwt(void);


/* DWT Variables */
#define DWT_CYCCNTENA_BIT (1UL << 0)
#define TRCENA_BIT (1UL << 24)


#define DWT_CONTROL (*((volatile uint32_t*) 0xE0001000))
#define DWT_CYCCNT (*((volatile uint32_t*) 0xE0001004))
#define DWT_LAR (*((volatile uint32_t*) 0xE0001FB0))
#define DEMCR (*((volatile uint32_t*) 0xE000EDFC))


#define DWT_CPICNT (*((volatile uint32_t* )0xE0001008))
#define DWT_EXCCNT (*((volatile uint32_t*) 0xE000100C))
#define DWT_SLEEPCNT (*((volatile uint32_t*) 0xE0001010))
#define DWT_LSUCNT (*((volatile uint32_t*) 0xE0001014))
#define DWT_FOLDCNT (*((volatile uint32_t*) 0xE0001018))
```

The dwt.h file contains the location of the peripheral registers in the Cortex-M memory map and constants to use for programming, as well as the function prototypes to start and stop the counter.

The dwt.c file, also available in the book's GitHub project, contains a function to unlock DWT and start counting, and another to stop counting and print the six counted values:

```
#include <stdio.h>
#include "dwt.h"

void start_dwt()
{
    DWT_LAR = 0xC5ACCE55;
    DEMCR |= TRCENA_BIT;
    DWT_CYCCNT = 0;
    DWT_CONTROL |= DWT_CYCCNTENA_BIT;
```

```
}

void stop_dwt()
{
    DWT_CONTROL &= ~DWT_CYCCNTENA_BIT;

    printf("CCNT = %u\n", DWT_CYCCNT);
    printf("CPICNT = %u\n", DWT_CPICNT);
    printf("EXCCNT  = %u\n", DWT_EXCCNT);
    printf("SLEEPCNT = %u\n", DWT_SLEEPCNT);
    printf("LSUCNT = %u\n", DWT_LSUCNT);
    printf("FOLDCNT = %u\n", DWT_FOLDCNT);
}
```

Here is another simple C program to measure the number of cycles a `printf` statement takes, this time using DWT counters:

```
#include <stdio.h>
#include "dwt.h"

int main()
{
    (void) start_dwt();

    printf("Count this hello world using DWT\n");

    (void) stop_dwt();
}
```

In the preceding SysTick and DWT examples, we covered only the minimal registers and programming values needed to effectively measure the performance of software on Cortex-M devices. CMSIS-Core includes a full description of the register interfaces and simplifies software reuse for more robust projects. Visit the CMSIS documentation for more information: `https://arm-software.github.io/CMSIS_5/General/html/index.html`.

If your Arm Cortex-M has a DWT, we recommend using the cycle counter to measure the cycles spent executing code. For Cortex-M processors without the **cycle counter (CYCCNT)** functionality of DWT, SysTick is a quality alternative to count cycles when there is no DWT available.

Measuring dot product performance

Now that we know how to measure the cycle count of an important section of code, let's give it a try by measuring the dot product performance on the Raspberry Pi Pico. We will look at multiple implementations of the dot product and experiment with compiler optimizations to see how the implementation of the dot product impacts performance.

Using the Raspberry Pi Pico

Often, a project already has a Cortex-M microcontroller chosen, which cannot be changed. In this case, the best system performance can be obtained using a combination of changes to the source code algorithms and the compiler optimization levels. In some cases, the compiler itself can also be changed (though this is often predetermined for projects).

In this section, we take the dot product example and create three different implementations with different source code and then use the compiler options to check the impact on performance. As the Cortex-M0+ in the Raspberry Pi Pico does not support CYCCNT DWT, we use the SysTick code provided in the previous section to count the clock cycles.

To replicate the example in this section, you will be using the following tools and environment:

Platform	Raspberry Pi Pico
Software	Dot Product
Environment	Raspberry Pi 4
Host OS	Linux (Ubuntu)
Compiler	GCC
IDE	-

For this example, we use the same C/C++ SDK introduced in *Chapter 3, Selecting the Right tools*, for the Pico. It uses the GNU compiler.

First, clone the project from the book's GitHub project by cloning the full project then navigating into the correct directory:

```
$ git clone  https://github.com/PacktPublishing/The-Insiders-
Guide-to-Arm-Cortex-M-Development.git
$ cd The-Insiders-Guide-to-Arm-Cortex-M-Development/chapter-5/
dotprod-pico
```

The C main function is in dotprod.c and defines two vectors each with 256 entries. There are three implementations to compute the dot product of these two vectors:

- dot_product1(): This is what most programmers would initially do. Make a loop, multiply, and add the vectors:

```
for (int i = 0; i < length; i++)
sum += v1[i] * v2[i];
```

- dot_product2(): This is a reasonable next step to optimize the calculation after discovering CMSIS-DSP and looking at the example programs. It uses a CMSIS-DSP function to multiply and another to add:

```
arm_mult_f32(v1, v2, multOutput, MAX_BLOCKSIZE);
for(i=0; i< MAX_BLOCKSIZE; i++)
arm_add_f32(&testOutput, &multOutput[i], &testOutput, 1);
```

- dot_product3(): This leverages the single dot product function in the CMSIS-DSP library:

```
(void) arm_dot_prod_f32(v1, v2, length, &result);
```

Look over each implementation and think about factors that impact performance. All implementations give the same numerical result but compute it using different instructions.

Results

Running the application on the Raspberry Pi Pico with the default compiler settings gives the following cycle counts. Please note that the exact cycle count may vary based on the compiler version:

Implementation	Number of cycles
1: Plain C code	41,668
2: CMSIS-DSP for multiply then add	54,539
3: CMSIS-DSP for dot product	41,808

Table 5.1 – Dot product performance, default optimization settings

Implementations 1 and 3 are very similar, and implementation 2 is slower because it has more function calls.

Let's take a look at how to work with the compiler flags.

There are four different settings for the GCC compiler. Each one can be specified when running `cmake`:

```
$ cmake -DCMAKE_BUILD_TYPE=<type>
```

The build type values are listed in the following table:

CMAKE_BUILD_TYPE	GCC flags used for optimization
Release	-O3 (max optimization)
Debug	-Og and -g (max debug)
RelwithDebInfo	-O2 and -g (good optimization with debug)
MinSizeRel	-Os (minimum code size)

Table 5.2 – Common GCC optimization flags

To see the actual compiler output and the flags used, set the VERBOSE flag for `make`. This will print the full compiler commands, and you can inspect the flags:

```
$ make VERBOSE=1
```

The VERBOSE build also reveals that the other flags are `-mcpu=cortex-m0plus -mthumb`. This is expected for Cortex-M0+.

During application creation and debugging, use the `Debug` target. This makes it easier for your debugger to step through code properly and makes improving code interactively much easier:

```
$ cmake -DCMAKE_BUILD_TYPE=Debug
```

The `Debug` build uses `-g` and `-Og` to optimize for debugging. The default `cmake` uses `-O3`. This will enable the majority of GCC optimizations. This is the `release` build type.

There is also `RelWithDebInfo` for the `release` build with debug info, which is `-O2` and `-g`. The final build type is `MinSizeRel`, which uses `-Os` to optimize for the smallest code size.

If you wish, you can override the values for each of the four build types. For example, to change the optimization level to `-O1` for the `release` build type, add the following line to the `CMakeLists.txt` file:

```
set(CMAKE_C_FLAGS_RELEASE "-O1 -DNDEBUG")
```

Then when you run `cmake` with the `release` build type, the `-O1` flag will be used instead of `-O3`:

```
$ cmake -DCMAKE_BUILD_TYPE=Release
```

Note that the variable is different for each build type. To change the optimization level of the Debug build type, set the CMAKE_C_FLAGS_DEBUG variable instead. Take some time and experiment with the compiler optimization levels to see different algorithm performances for each type of dot product implementation.

For a minimum-size build type, the cycle counts are in the following table. As mentioned earlier, you might notice different cycle counts based on the compiler version you use:

Implementation	Number of cycles
1: Plain C code	42,696
2: CMSIS-DSP for multiply then add	56,902
3: CMSIS-DSP for dot product	42,937

Table 5.3 – Dot product performance, minimum size optimization settings

In this case, the minimum size build resulted in worse performance than the Debug build for each dot product implementation.

There are other changes to make besides the compiler flags to impact performance. The CMSIS functions have alternative implementations for loop unrolling. These can be seen in the source files such as arm_dot_prod_f32.c. To review, open the source file with a text editor and look for this line:

```
#if defined (ARM_MATH_LOOPUNROLL) && !defined(ARM_MATH_
AUTOVECTORIZE)
```

Edit CMakeLists.txt to set the value for loop unrolling and run again to see the performance impact:

```
set(CMAKE_C_FLAGS_RELEASE "-O3 -DNDEBUG -DARM_MATH_LOOPUNROLL")
```

The new values are listed in the following table:

Implementation	Number of cycles
1: Plain C code	41,668
2: CMSIS-DSP for multiply then add	60,887
3: CMSIS-DSP for dot product	41,034

Table 5.4 – Dot product performance, with loop unrolling

The plain C code implementation runs the same as the previous -O3 implementation as it is not a CMSIS function. The second implementation increased in cycles, while the third implementation decreased and thus had improved performance. This reveals a truth about optimizing performance for a multi-faceted problem such as embedded software development: it is both a science and an art.

Making informed decisions about what Cortex-M processor, algorithm implementation, and compiler options to select is crucial to optimizing performance. It is equally important to modify some variables and measure the results yourself for your unique situation. The interplay between compilers, compiler flags, hardware, and software is a complex domain where experience is the best teacher.

To gain a deeper understanding of system behavior, you should always look at the disassembly file to see the actual code generated. For example, the dotprod.dis file generated in the build directory shows the difference in code with and without the loop unrolling. If you have questions about why a certain implementation has a different performance than expected, it helps to be able to look at and generally understand the assembly code.

The next section will take the same three dot product implementations and try them on different hardware using a different compiler.

Using NXP LPC55S69-EVK

This section ports over the same software that ran dot products on the Pico to run on the NXP LPC55S69-EVK board. In this case, the Cortex-M33 supports DWT and will use that to measure cycles here as opposed to the SysTick counter.

To replicate the example in this section, you will be using the following tools and environment:

Platform	NXP LPC55S69-EVK
Software	Dot Product
Environment	Personal Computer
Host OS	Windows
Compiler	Arm Compiler for Embedded
IDE	Keil MDK Community

While you can create a new project in Keil µVision to build and run the dot product example, the easiest way is to modify the existing hello_world demo example. The steps to download and install the hello_world example using Pack Installer are the same as the ones you would have followed in *Chapter 4, Booting to Main*, while running the led_blinky example. The instructions are already documented here: https://developer.arm.com/documentation/kan322/latest/.

Once you've loaded the hello_world project in μVision, replace the contents of the `hello_world.c` source file with the file containing our dot product algorithms, `dot_product.c`. You can find it at our book's GitHub link: `https://github.com/PacktPublishing/The-Insiders-Guide-to-Arm-Cortex-M-Development/tree/main/chapter-5/dotprod-nxp-lpcxpresso55s69/hello_world.c`. Make sure to use the correct `dot_product.c` file intended for the NXP board, as it has code specific to its memory map. You can also ignore the `dwt.c` and `dwt.h` files in that GitHub repository, as they are automatically included in the example hello_world project through CMSIS and are there only for reference.

Then *save* and *build* your project with the dot product algorithms in the `hello_world.c` file. This project uses the `arm_math.h` file, so you must include the CMSIS-DSP library in your build. To do so, select **Manage Run-Time Environment** from the top of the IDE's GUI and check the box under **CMSIS | DSP**:

Software Component	Sel.	Variant	Version	Description
⊞ ❖ Board Support				Generic Interfaces for Evaluation and Development Boards
⊟ ❖ CMSIS				Cortex Microcontroller Software Interface Components
● CORE	☑		5.6.0	CMSIS-CORE for Cortex-M, SC000, SC300, Star-MC1, ARMv8-M, ARMv8.1-M
● DSP	☑	Source	1.10.0	CMSIS-DSP Library for Cortex-M, SC000, and SC300
● NN Lib	☐		3.1.0	CMSIS-NN Neural Network Library
⊞ ❖ RTOS (API)			1.0.0	CMSIS-RTOS API for Cortex-M, SC000, and SC300
⊞ ❖ RTOS2 (API)			2.1.3	CMSIS-RTOS API for Cortex-M, SC000, and SC300
⊞ ❖ CMSIS Driver				NXP MCUXpresso SDK Peripheral CMSIS Drivers
⊞ ❖ Compiler		ARM Compiler	1.7.2	Compiler Extensions for ARM Compiler 5 and ARM Compiler 6
⊞ ❖ Device				Startup, System Setup
⊞ ❖ File System		MDK-Plus	6.15.0	File Access on various storage devices
⊞ ❖ Graphics		MDK-Plus	6.24.0	User Interface on graphical LCD displays
⊞ ❖ Network		MDK-Plus	7.17.0	IPv4 Networking using Ethernet or Serial protocols
⊞ ❖ USB		MDK-Plus	6.16.0	USB Communication with various device classes

Figure 5.1 – Adding CMSIS-DSP library in Keil MDK

Once you are able to build your project successfully, you can choose between two different configurations for building your project: debug and release. The compiler and linker optimization settings differ based on your selection. The main difference in the default settings is that *debug* uses `-O1` and *release* uses `-Oz`.

These are the compiler settings for the Debug build:

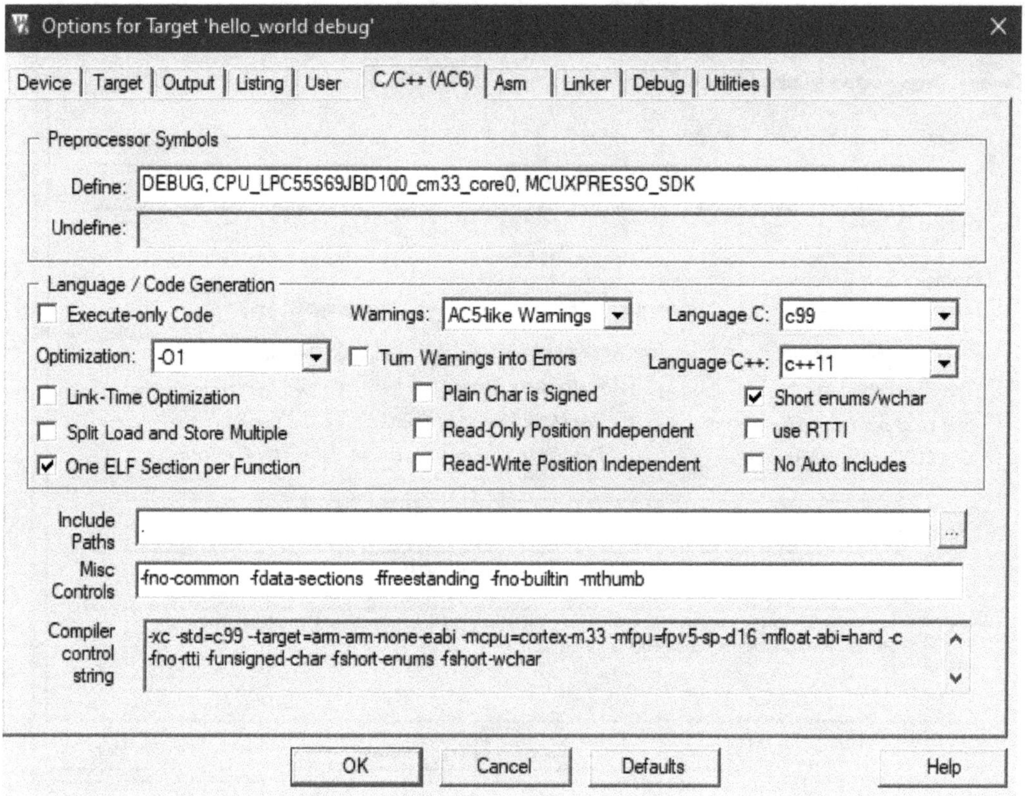

Figure 5.2 – Debug build compiler settings

These are the compiler settings for the `Release` build:

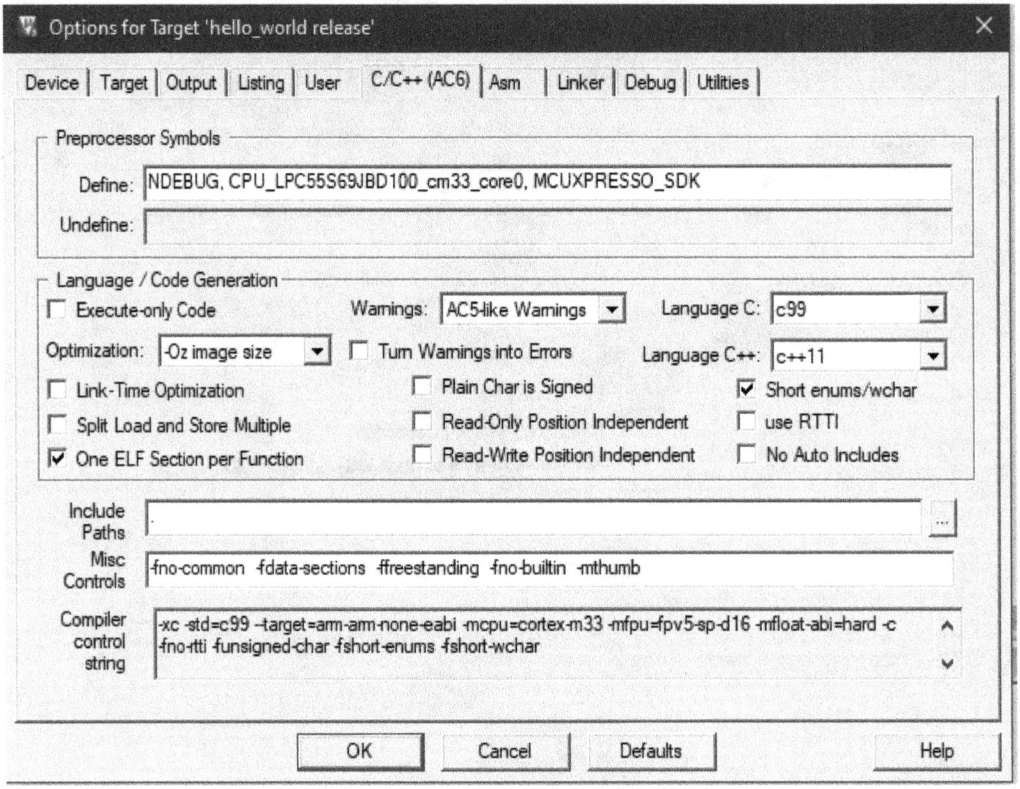

Figure 5.3 – Release build compiler settings

Similar to GCC, Arm Compiler for Embedded has specific compiler optimization settings for different situations. They can be summarized as follows:

Ideal for	Arm Compiler for Embedded flags for optimization
Performance	`-Ofast or -O3`
Debug, low performance	`-O0`
Debug, more optimized	`-O1`
Minimum code size	`-Oz`

Table 5.5 – Common Arm Compiler for Embedded optimization flags

With this understanding, we can now try different compiler settings and see the resulting performance.

For optimization levels above -O0, the compiler will inline the start_dwt() and stop_dwt() functions. This may result in the unexpected ordering of the functions, and the cycle count of the dot product is not actually measured because stop_dwt() happens before the calculation is done.

Please add -fno-inline-functions to the **Misc Controls** box on the **Options for Target** screen as shown in *Figure 5.3*.

Results

When running with the debug option with the flag set to -O1, these are the resulting numbers:

Implementation	Number of cycles
1: Plain C code	2,398
2: CMSIS-DSP for multiply then add	9,765
3: CMSIS-DSP for dot product	2,422
Minimum code size	-Oz

Table 5.6 – Dot product performance, debug optimization settings

Note that when clicking **build** in the Keil IDE, it also reports the code size for the software. This is another axis to measure optimization, not on performance but on code size. The code size for the preceding debug option is 6,700 bytes, which can vary based on compiler versions.

Running with the release option with the flag set to -Oz, these are the resulting numbers:

Implementation	Number of cycles
1: Plain C code	2,100
2: CMSIS-DSP for multiply then add	8,891
3: CMSIS-DSP for dot product	1,857

Table 5.7 – Dot product performance, minimum size optimization settings

This has a code size of 5,232, which is lower than expected.

Lastly, change the compiler optimization flag to the maximum performance `-Ofast` and see the result. To do this, go into **Project | Options for Target 'hello_world release' |** the **C/C++(AC6)** tab | **Optimization**. Set the `Optimization` flag to `-Ofast`. Here are the results:

Implementation	Number of cycles
1: Plain C code	1,752
2: CMSIS-DSP for multiply then add	11,246
3: CMSIS-DSP for dot product	1,652

Table 5.8 – Dot product performance, fast optimization settings

This has a code size of 7,944.

As with the Pico, some performance gains are expected, and others respond in less intuitive ways. This is why trying different compiler options and algorithm implementations leads to better performance over time. The next section will cover the same topic for the Cortex-M55.

Using Arm Virtual Hardware

This section measures the same dot product code, slightly modified to run on the Arm Virtual Hardware Cortex-M55 FVP. It measures the cycle count using the SysTick timer.

To replicate the example in this section, you will be using the following tools and environment:

Platform	Arm Virtual Hardware – Corstone-300
Software	Dot Product
Environment	Amazon EC2 (AWS account required)
Host OS	Ubuntu Linux
Compiler	Arm Compiler for Embedded
IDE	-

The setup for this example is largely the same as in *Chapter 4, Booting to Main*. Set up a new AMI of Arm Virtual Hardware on Amazon EC2, download the example via the GitHub link for this example, build, and run. These are the commands to run on your AMI cloud instance:

```
git clone  https://github.com/PacktPublishing/The-Insiders-
Guide-to-Arm-Cortex-M-Development.git
    cd  The-Insiders-Guide-to-Arm-Cortex-M-Development/
chapter-5/dotprod-avh-corstone-300
```

```
    ./build.sh
    ./run.sh
```

Results

You will then get these results with the default Arm Compiler for Embedded flags:

Implementation	Number of cycles
1: Plain C code	4,294,967,295
2: CMSIS-DSP for multiply then add	11,390,997
3: CMSIS-DSP for dot product	5,201,475

Table 5.9 – Dot product performance, default optimization settings

If these results seem so high as to be incorrect, your intuition is right! Our method of measurement in the previous sections has been using DWT and SysTick, which depend on physical oscillating counters on the Pico and NXP boards to record accurate cycle counts. On this virtual Corstone-300, however, there are no physical oscillating counters to read from; the SysTick in this example is abstracted to work at the functional level, but not at a cycle-accurate level.

This *functional accuracy* of software gives Arm Virtual Hardware a large advantage when you just need software to behave as it would in the end system without cycle accuracy. You can spin up hundreds of virtual boards to run CI/CD tests in parallel without buying as many boards (a topic covered in *Chapter 9, Implementing Continuous Integration*). You can easily start working with Arm hardware without buying physical hardware. These advantages are possible because the virtual hardware runs as fast or faster than the same software running on a physical Cortex-M board. Simulating software at the cycle-accurate level requires a significant amount of complexity, which slows down execution to the point of being useless for these software development use cases. This is the trade-off of functional accuracy versus cycle accuracy.

What this means in this chapter's context is that it is generally not helpful to measure and optimize software performance on virtual hardware. The closest you can get is a relative understanding of performance. In this example, using plain C is an order of magnitude slower than CMSIS-DSP for multiply and add, which is an order of magnitude slower still than using the CMSIS-DSP dot product function. This indicates that implementation 3 is better than 2, which is better than 1, but you should always verify on physical hardware to obtain realistic measurements.

Optimization takeaways

We have evaluated the performance of the dot product while altering the following variables:

- Processor type
- Software source code
- Compiler and compiler options

To provide some helpful guidelines when optimizing a Cortex-M system, here is a summary table of the recorded cycle counts when altering the dot product algorithm and compiler flags across both the Pico and NXP boards. As described in the previous section, the Corstone-300 Arm Virtual Hardware system does not allow cycle-accurate measurements, so we will not include it in our summary table here:

		RPi	NXP
Implementation	**Compiler Flags**	**M0+**	**M33**
1: Plain C code	Debug	41,668	2,398
	Release Size	42,696	2,100
	Release Speed	41,668	1,752
2: CMSIS-DSP for multiply then add	Debug	54,539	9,765
	Release Size	56,902	8,891
	Release Speed	60,887	11,246
3: CMSIS-DSP for dot product	Debug	41,808	2,422
	Release Size	42,937	1,857
	Release Speed	41,034	1,652

Table 5.10 – Comparing optimization techniques for dot product

Looking at the preceding table, we can make some specific observations, as well as some helpful generalizations about Cortex-M optimization.

Processor performance

It is clear from the results that the NXP board, based on the Cortex-M33, runs every dot product implementation at every compiler optimization level faster than the Pico, based on the Cortex-M0+. This should not be surprising as the Cortex-M33 is designed to be much more powerful than the Cortex-M0+ in mathematical operations, especially as it includes the FPU.

Importantly, the difference in performance will be more or less prevalent with different software. Running a more realistic software stack and complex algorithms will likely differentiate the Cortex-M33 from the Cortex-M0+ even more. Ultimately, the delta between processors will vary on how your specific software leverages the capabilities of each. And faster is not always better; if you are optimizing for cost and a Cortex-M0+ can run your algorithm fast enough, that may be the wiser choice.

Compilers

In this chapter, we used GCC for the Raspberry Pi Pico and Arm Compiler for Embedded for the NXP LPC55S69-EVK. While the difference in performance is clearly attributable to the Cortex-M33 capabilities over the Cortex-M0+, in general, the Arm Compiler for Embedded has a performance edge over the open source GCC. This is because the Arm Compiler for Embedded is developed alongside the Cortex-M processors, and it is explicitly designed to extract as much performance from each processor as possible. Over time, most of these performance tricks are upstreamed into the GNU compiler, but this can take years after a processor is released. The performance difference is more prominent for newer Cortex-M processors but is generally present for all of them.

In use cases where maximizing application performance is tantamount, this compiler performance difference is another variable to improve. For most cases in the Cortex-M space, however, using GCC or Arm Compiler for Embedded (or another compiler) will not make or break your application.

Compiler flags

The results of altering the flags are a bit more mixed. It is really important to understand what compiler flags can be changed in different settings. This chapter gave examples of the optimization flag and the loop unrolling flag, and there are other optimizations to explore to ensure you are extracting all the performance possible from your system. When using Arm Compiler for Embedded for more realistic software applications, investigate **Link Time Optimization** (**LTO**). This is a powerful technique for pushing the performance of applications and can induce unexpected software behavior and cause runtime errors when not implemented correctly.

Source code / algorithm implementation

The key takeaway is to use pre-optimized algorithms when available. Leveraging the CMSIS dot product consistently gave better performance than the other two implementations. In general, CMSIS libraries offer highly optimized implementations of common algorithms; where possible, use them to your advantage.

For this simple dot product example, it was mostly a worse implementation to use two CMSIS functions (multiply then add) as opposed to a plain C implementation due to the overhead caused by transferring data between functions. As your software gets more complex, this type of overhead versus efficiency trade-off may prove worth it.

In general, that overhead versus efficiency trade-off can be quite complex in a realistic software setting. It is up to you to explore different optimization options with your given constraints on time, cost, hardware, and software. Take the lessons from this chapter to direct your focus and make educated decisions about altering your project to optimize it well.

Summary

This chapter outlined basic principles of performance optimization for Cortex-M systems. We took an example software algorithm, the dot product, and went through several boards and compiler options to explore performance implications. The skills learned in this chapter will help you optimize your Cortex-M system intelligently by altering the main factors influencing performance: processor type, software source code implementation, and compilers and compiler options.

No matter how optimized your system is for performance, there are still other important aspects to consider that ensure you have a quality Cortex-M product. The next chapter provides both an overview and practical guide for using machine learning on edge devices.

Further reading

- **Link Time Optimization (LTO)** overview, with implementation examples and potential side-effects: https://developer.arm.com/documentation/100748/latest/Writing-Optimized-Code/Optimizing-across-modules-with-Link-Time-Optimization

- Arm Compiler for Embedded optimization flag options reference and guide: https://developer.arm.com/documentation/100748/latest/Using-Common-Compiler-Options/Selecting-optimization-options?lang=en

- GCC optimization flag options reference and guide: https://gcc.gnu.org/onlinedocs/gcc/Optimize-Options.html

- SysTick functional description and register reference, included in the Cortex-M33 technical reference manual: https://developer.arm.com/documentation/dui0662/b/Cortex-M0--Peripherals/System-timer--SysTick

- **Data Watchpoint and Trace (DWT)** functional description and register reference, included in the Cortex-M33 technical reference manual: https://developer.arm.com/documentation/100230/0002/debug-and-trace-components/data-watchpoint-and-trace-unit/dwt-functional-description

- CMSIS-DSP documentation to learn more about these hand-optimized libraries, freely available and highly recommended for Cortex-M developers: https://www.keil.com/pack/doc/CMSIS/DSP/html/modules.html

- A guide to writing optimized code with Arm Compiler for Embedded, containing very specific optimization techniques: https://developer.arm.com/documentation/100748/0618/Writing-Optimized-Code

6
Leveraging Machine Learning

ML applications have grown to dominate in highly visible and enterprise-scale uses today: Google search results, Facebook/Instagram/TikTok/Twitter sorting algorithms, YouTube's suggested content, Alexa/Siri voice assistants, internet advertising, and more. These use cases all host their ML models and perform their inferences in the cloud, then show the results to us as end users on our edge devices such as phones, tablets, or smart speakers. This paradigm is beginning to change, with more models being stored (and inferences being run) on the edge devices themselves. The shift to processing at the edge removes the need for transmission to and storage in the cloud, and as a result, provides the benefits listed here:

- **Enhanced security**: Reducing attack vectors
- **Enhanced privacy**: Reducing the sharing of data
- **Enhanced performance**: Reducing application latency

This chapter is intended to give an overview and practical guide for using ML on edge devices. It will cover the following topics:

- The high-level life cycle of creating and maintaining an ML application
- Key ML libraries/software to be aware of
- Three hands-on examples using different ML algorithms on embedded platforms

Understanding the ML application life cycle

The *life cycle* to create, maintain, and update an ML application can be visualized in a few different ways. The following diagram shows a high-level view of this ML life cycle for a typical ML project being deployed on edge devices. Take a minute to look through the listed steps before we explain each section:

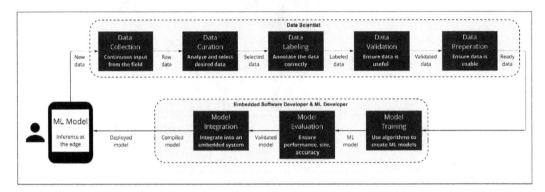

Figure 6.1 – Typical ML life cycle for embedded projects

The steps at the top all pertain to data. Gathering it and preparing it for use in ML algorithms is a non-trivial task and is often the source of competitive advantage for companies (as opposed to the ML algorithms themselves). Commonly, there is a dedicated resource—or resources—for preparing data at companies, referred to here as data scientists. Once the data is prepared, the data is used to create and train an ML model. This is commonly performed by an embedded software developer and—potentially—an ML developer that specializes in ML models and training. The data scientist will likely help with model evaluation as well.

The bottom three steps are of most interest to us in this chapter, as data preparation and compiled software deployment are outside the scope of this book. These three steps are model training, model evaluation, and model integration.

The *model training phase* takes prepared data and builds an ML model for use in the embedded device. An algorithm must be selected that meets your goals and is possible to implement with the data you possess. Algorithm selection is another detailed topic in and of itself. There is an excellent algorithm cheat sheet from Microsoft on this topic. Refer to it for further guidance here: `https://docs.microsoft.com/en-us/azure/machine-learning/media/algorithm-cheat-sheet/machine-learning-algorithm-cheat-sheet.png`.

While we are not going into deep detail, here is a brief guide to algorithm types based on what your goals are:

- Extract information from text: "What information is in this text?"

  - Text-analysis algorithms are best, such as feature hashing and word2vec

- Predict future values: "How much of this will there be?"

  - Regression algorithms are ideal, including linear regression or **neural network** (**NN**) regression

- Find unusual occurrences: "Is this unusual?"

  - Anomaly detection algorithms, such as one-class **Support Vector Machine** (**SVM**)

- Discovering structure: "How does this have structure?"

  - Clustering algorithms, such as the unsupervised K-Means algorithm

- Generate recommendations: "What would this person like to see?"

  - Recommendation algorithms, such as **Singular Value Decomposition** (**SVD**)

- Classify images: "What is this image?"

  - Image classification algorithms, such as ResNet or DenseNet

- Predict between categories: "What is this thing most like?"

  - Classification, either two-class or multi-class algorithms such as two-class NNs or multi-class decision forest

The terms **supervised learning** (**SL**) and **unsupervised learning** (**UL**) refer to the nature of how the ML algorithm learns from data. Put simply, if data is labeled, then you can use an SL algorithm (as you are supervising the learning by labeling the data and verifying its correct analysis during model training). If data is not labeled, then you use UL algorithms. This is ideal for discovering hidden patterns in data, as the ML algorithm is not bound or influenced by defined labels.

The *model evaluation phase* is for you to ensure the model has been trained to your specifications. The most common metrics for success in embedded systems are noted here:

- **Accuracy**: Critical to the application's success is the ability to provide sensible results. The margin for errors is specific to your use case and industry and can be evaluated through appropriate training versus test dataset splits.

- **Performance**: The inference can run within a certain time to ensure application usability in practice. This performance should be measured from the inference running on the edge device (or a simulation of the edge device) to give realistic results. Running tests on your actual device, similar development boards, or even **Arm Virtual Hardware** (**AVH**) can help give pointers on inference performance.

- **Size**: The model must fit on your edge device, which has a limited amount of memory that can hold the model. This can be measured after the model is trained and compiled for your specific system.

The *model integration phase* is when you take the model and integrate it within your other application code. At this point, you turn the ML model into a C++ (or similar) file and stitch it into the overall system through the typical *code, compile, run, debug* cycle of embedded development.

Finally, after the model is integrated into your overall application that has been compiled, it can be deployed to your edge device. This may involve updating an existing device with a new model or creating a new device to be distributed. In either case, at this point, your model is now deployed and being used.

With this high-level overview of the life cycle of an ML model in embedded systems, it is time to look at some specific libraries that provide algorithms and stitching capabilities that simplify this process.

Investigating ML frameworks and libraries

The list of frameworks and libraries that help bring ML to edge devices is large and constantly evolving. This section will highlight the top three most common ML frameworks and libraries for Cortex-M ML at the time of writing this book. We will cover the following:

- TensorFlow Lite for Microcontrollers
- CMSIS-NN
- TinyML

TensorFlow Lite for Microcontrollers

The first framework to discuss is **TensorFlow Lite for Microcontrollers (TFlite Micro)**. TensorFlow Lite is a framework that is designed for mobile devices, microcontrollers, and other edge devices where a small memory footprint and optimized performance are important. TFlite Micro is a C++ framework for ML inference that works well on Cortex-M microcontrollers. It is an optimized version of TensorFlow for embedded C/C++ applications. It is designed to run on 32-bit microcontrollers and results in very small binary files. It also doesn't require an **operating system (OS)**, standard C++ libraries, or dynamic memory allocation functions. A small memory footprint combined with the power of TensorFlow makes TFlite Micro one of the most popular ML frameworks for **Internet of Things (IoT)** applications.

Note that at present, this is one of the only fundamental frameworks for bringing ML to Cortex-M devices, so you will almost certainly work with this framework if introducing ML capabilities to your edge device. In terms of the life cycle of ML projects detailed earlier, the data selection and normalization—as well as model training and testing—will be performed through TensorFlow Lite on a desktop computer (or cloud environment). In the model integration step, you will then convert this trained model into a TFlite Micro model, then finally into your embedded application itself.

CMSIS-NN

The **Common Microcontroller Software Interface Standard** (**CMSIS**) and CMSIS-NN were previously mentioned in *Chapter 1, Selecting the Right Hardware*. CMSIS-NN is tightly integrated with TFlite Micro providing optimized versions of TFlite Micro kernels in an easy-to-understand way. CMSIS-NN is an excellent choice for bringing ML capabilities to Cortex-M devices in an optimized fashion without rewriting existing functionality.

TinyML

TinyML is relatively new on the scene, first introduced in 2018 and popularized by Google's Pete Warden in his book on the topic in 2020. It is primarily a concept of generally bringing ML to small IoT devices as opposed to a specific library. TinyML is a useful term to be aware of as it may refer to projects that leverage TFlite Micro or other languages such as MicroPython. An example of TinyML in practice using MicroPython on a Cortex-M4 device can be found in the *Further reading* section at the end of this chapter.

Now that we have looked at the commonly used ML frameworks and libraries for the development of your ML applications on Cortex-M-based devices, let us dive into exploring some of these applications with hands-on examples.

Exploring ML use cases

ML is rapidly evolving to more use cases such as fraud detection, personalized targeted marketing, and self-driving cars, to name a few. Currently, the most popular applications for ML in the embedded space can be grouped into three areas: vibration, vision, and voice. These categories were briefly discussed in *Chapter 1, Selecting the Right Hardware*, with the current chapter focusing on running examples in each of these areas. Vibration typically is the most straightforward and is therefore a good starting point to begin exploring. After familiarizing ourselves with the vibration example, we provide more details on the construction of the ML software stack in the vision and voice examples that follow. The three use cases are presented here:

- Anomaly detection using the Cortex-M55

- Image classification using the Cortex-M55 and Ethos-U55

- Micro speech using the Cortex-M55 and Ethos-U55

Anomaly detection – vibration

To replicate the example in this section, you will be using the following tools and environments:

Platform	Arm Virtual Hardware – Corstone-300
Software	ML Application (Anomaly detection)
Environment	Amazon Elastic Compute Cloud (EC2) (Amazon Web Services (AWS) account required)
Host OS	Ubuntu Linux
Compiler	Arm Compiler for Embedded
IDE	-

Vibration ML applications gather motion data, typically from an accelerometer, and analyze it periodically to ensure normal behavior. This type of analysis can be found in the health industry and particularly in ensuring industrial machines are working correctly. These algorithms are ideal to detect unusual mechanical movements that are precursors to failure, enabling companies to take preventative action before expensive machinery breaks down.

The example in this section is anomaly detection based on audio data that comes from the Arm ML Embedded Evaluation Kit. This kit is a great resource to get started and see how to run realistic ML workloads on an example system. The example system, in this case, is the Corstone-300 **Fixed Virtual Platform** (FVP) provided in AVH, enabling the exploration of these models without any physical hardware.

An anomaly detection example can be found here: `https://review.mlplatform.org/plugins/gitiles/ml/ethos-u/ml-embedded-evaluation-kit/+/HEAD/docs/use_cases/ad.md`. It illustrates how an anomaly detection flow could be used to monitor for anomalies in running industrial machines. This software example uses audio recordings instead of vibration from an accelerometer, but the principle is the same between the two data types. The software records an audio sample, performing inferences to identify whether the audio sample is outside a set acceptable threshold range of variance from the normal. If the sample is outside this threshold, the machine is flagged as behaving anomalously and corrective action could be taken.

Environment setup

To get started, launch the AVH **Amazon Machine Image** (AMI) in AWS. You will then need to connect to the AMI. The steps are documented here: `https://arm-software.github.io/AVH/main/infrastructure/html/run_ami_local.html#connect`.

Note that these examples use the Corstone-300 FVP in AVH display features to show images and serial output on a separate screen, so an X server is required to run properly. Connecting to your EC2 instance via the online AWS terminal, or a simple `ssh` via Command Prompt in Windows, will not properly forward an X display to run the example correctly. Make sure you connect to your EC2 instance with a method that enables display. Here are some options based on the OS you are using:

- Windows
  - MobaXterm
- Mac/Linux
  - XQuartz
- Full **graphical user interface (GUI)** options for any OS
  - NoMachine
  - Any **virtual network computing (VNC)** client

Connect to an AWS instance of the AMI using `ssh`. This was covered previously in *Chapter 4, Booting to Main*. The code is illustrated here:

```
$ ssh -i <your-private-key>.pem ubuntu@<your-ec2-IP-address>
```

Start by cloning the Git repository for the previously mentioned ML Embedded Evaluation Kit from Arm. It will automatically download several example ML use cases that run on the Corstone-300 FVP, including this example and the next section's example as well on image classification. The code is illustrated here:

```
git clone --recursive "https://review.mlplatform.org/ml/
ethos-u/ml-embedded-evaluation-kit"
```

Then, enter the directory to proceed with downloading all the files required to build this example:

```
cd ml-embedded-evaluation-kit/
```

Before building the example software, you need to download some dependencies and resources that are not included in the Git repository due to space and third-party arrangements. Enter this command first to download the CMSIS, TFlite, and Ethos-U55 drivers:

```
python download_dependencies.py
```

If this command fails with a **Dependencies folder exists. Skipping download** message, then delete the dependencies folder with `rm -rf dependencies/` and try again.

Lastly, you will need to set up the resources just downloaded from Git and the `download_dependencies.py` script. Do so with this command:

```
python set_up_default_resources.py
```

If this command fails, debug depending on the error you obtain, as follows:

- If it's a failure due to `resources_downloaded directory exists`, then delete this folder with `rm -rf resources_downloaded/` and try again

- If it doesn't work due to `python3 -m venv env` failing, then install the `python3` virtual environment with the `sudo apt install python3.8-venv` command, and try again

It will take about a minute to download, and then you are ready to build the example application.

Build

This repository uses `cmake` (`https://cmake.org/`) as the build system. The applications can be built using either the Arm Compiler for Embedded toolchain or the **GNU Compiler Collection** (**GCC**). This book will use the Arm Compiler for Embedded toolchain here, which is already included in the AMI and initialized in the previous steps.

Make a `build` directory, enter it, and run the following `cmake` command followed by `make` to kick off the build process. The exact commands are shown here:

```
mkdir cmake-build-mps3-sse-300-arm
cd cmake-build-mps3-sse-300-arm
cmake .. -DTARGET_PLATFORM=mps3 -DTARGET_SUBSYSTEM=sse-300
-DETHOS_U_NPU_ENABLED=OFF -DCMAKE_TOOLCHAIN_FILE=~/ml-embedded-
evaluation-kit/scripts/cmake/toolchains/bare-metal-armclang.
cmake -DUSE_CASE_BUILD=ad
make
```

This will take about 2 minutes or so to complete, with the final executable being built as a result: `ethos-u-ad.axf`. ad stands for **anomaly detection** here. Note that it is located in the `cmake-build-mps3-sse-300-arm/bin/` directory. The **neural processing unit** (**NPU**) is disabled, even with the name of the `.axf` file including `ethos-u`.

Before we move on to running this executable on the Corstone-300 FVP in AVH, let's look at some of the options we passed to the preceding `cmake` command, what they mean, and how we can change them for different results, as follows:

- `-DTARGET_PLATFORM=mps3 -DTARGET_SUBSYSTEM=sse-300`

  - These commands specify to use the MPS3 platform and SSE-300 subsystem as the targets for building. This will build the application for the correct memory map of the VHT_Corstone_SSE-300_Ethos-U55 FVP, whether or not we take advantage of the Ethos-U55 accelerator.

- `-DCMAKE_TOOLCHAIN=~/ml-embedded-evaluation-kit/scripts/cmake/bare-metal-gcc.cmake`

  - This will set your compiler toolchain as GCC instead of the default Arm Compiler for Embedded.

- `-DETHOS_U_NPU_ENABLED=OFF`

  - This specifies that we are just running the ML application on the Cortex-M55 processor, not taking advantage of the Ethos-U55 acceleration. This simple vibration detection use case does not need acceleration, but the next example of image classification will show specifically how to use this in practice.

- `-DETHOSU_TARGET_NPU_CONFIG=H32`

  - Only applies when `-DETHOS_U_NPU_ENABLED=ON`, when using the NPU as an accelerator. This option defines the configuration of the NPU that you are building the application for. It corresponds to the number of 8x8 **multiply-accumulates** (**MACs**) per cycle performed by the NPU. The allowed values are H32, H64, H128, and Y256. H128 is the default setting, and you can override it using this parameter—for example, to build for 32 MACs.

Now that we have looked at the steps to build the anomaly detection application, let us move on to running the executable.

Run

To execute the anomaly detection example we built, execute the following command:

```
VHT_Corstone_SSE-300_Ethos-U55 -a ~/ml-embedded-evaluation-kit/
cmake-build-mps3-sse-300-arm/bin/ethos-u-ad.axf
```

`-a` points to the application that we are running on the model.

Note again that if you get an error about `Display not set` or an `xterm` error, you need to ensure your SSH connection method has X-forwarding and your system has an X server. Refer to the start of the example to fix this problem. You may also see a `Warning: Failed to write bytes at address range` ... message pop up, but that is just a warning, and it is running just fine.

A **liquid-crystal display** (**LCD**) terminal will appear, replicating the MPS3 physical board's LCD board behavior. Then, another Telnet terminal will also appear, displaying the program output and inference status, as depicted in the following screenshot:

```
INFO - Application Note AN228, Revision C
INFO - MPS3 build 3
INFO - MPS3 core clock has been set to: 32000000Hz
INFO - CPU ID: 0x411fd220
INFO - CPU: Cortex-M55 r1p0

INFO - Enabling I-cache.
INFO - Enabling D-cache.
INFO - Target system design: Arm Corstone-300 - AN552
INFO - ARM ML Embedded Evaluation Kit
INFO - Version 22.08.0 Build date: Oct  4 2022 @ 18:08:44
INFO - Copyright (C) ARM Ltd 2021-2022. All rights reserved.

INFO - Added  support to op resolver
INFO - Creating allocator using tensor arena at 0x31000000
INFO - Allocating tensors
INFO - Model INPUT tensors:
INFO -   tensor type is INT8
INFO -   tensor occupies 1024 bytes with dimensions
INFO -          0:   1
INFO -          1:  32
INFO -          2:  32
INFO -          3:   1
INFO - Quant dimension: 0
INFO - Scale[0] = 0.192437
INFO - ZeroPoint[0] = 11
INFO - Model OUTPUT tensors:
INFO -   tensor type is INT8
INFO -   tensor occupies 8 bytes with dimensions
INFO -          0:   1
INFO -          1:   8
INFO - Quant dimension: 0
INFO - Scale[0] = 0.048891
INFO - ZeroPoint[0] = -30
INFO - Activation buffer (a.k.a tensor arena) size used: 275660
INFO - Number of operators: 14
INFO -   Operator 0: CONV_2D
INFO -   Operator 1: DEPTHWISE_CONV_2D
INFO -   Operator 2: CONV_2D
INFO -   Operator 3: DEPTHWISE_CONV_2D
INFO -   Operator 4: CONV_2D
INFO -   Operator 5: DEPTHWISE_CONV_2D
INFO -   Operator 6: CONV_2D
INFO -   Operator 7: DEPTHWISE_CONV_2D
INFO -   Operator 8: CONV_2D
INFO -   Operator 9: DEPTHWISE_CONV_2D
INFO -   Operator 10: CONV_2D
INFO -   Operator 11: AVERAGE_POOL_2D
INFO -   Operator 12: CONV_2D
INFO -   Operator 13: RESHAPE
INFO - Running inference on audio clip 0 => random_id_00_000000.wav
INFO - Inference 1/1
INFO - Average anomaly score is: -0.883147
INFO - Anomaly threshold is: -0.800000
INFO - Everything fine, no anomaly detected!
INFO - Main loop terminated.
INFO - program terminating...
```

Figure 6.2 – Anomaly detection output

This execution will take about 2 minutes and will finish by displaying a success inference and a non-anomalous machine. Running the ML model on the Ethos-U55 will significantly speed up the execution, as it is intended to.

This example was a quick-start guide to running an ML algorithm on a Cortex-M55-based system, using virtual hardware. The next example will show the different commands to enable and run an example on the Ethos-U55 included in the same system with the image classification algorithm.

Image classification – vision

To replicate the example in this section, you will be using the following tools and environments:

Platform	Arm Virtual Hardware – Corstone-300
Software	ML Application (Image classification)
Environment	Amazon EC2 (AWS account required)
Host OS	Ubuntu Linux
Compiler	Arm Compiler for Embedded
IDE	-

Vision ML applications can range greatly in uses. Object detection is used in self-driving cars, face unlocking is used in mobile devices and smart home cameras, and image classification is common across many industries for various purposes. The value of vision-based algorithms at the edge will only grow over the coming years.

Image classification refers to the task of identifying what an image represents. For example, you can train an NN model to recognize images of different animals such as a cat or a dog. This image classification example uses the MobileNetV2-1.0 quantized `uint8` NN model, which is openly available on TensorFlow Hub here: `https://tfhub.dev/tensorflow/lite-model/ mobilenet_v2_1.0_224_quantized/1/default/1`. The pre-trained network model is trained on more than a million images from the ImageNet database (`https://image-net. org/`) and can classify images into 1,000 object categories. If you are curious and want to look at all the object categories, take a look here: `https://github.com/google-coral/edgetpu/ blob/master/test_data/imagenet_labels.txt`.

The example in this section can be found here: `https://review.mlplatform.org/plugins/ gitiles/ml/ethos-u/ml-embedded-evaluation-kit/+/HEAD/docs/use_cases/ img_class.md`. It runs classification on four different input images (cat, dog, kimono, and tiger). Note that this is the same Arm ML Embedded Evaluation Kit from the previous section on anomaly detection and will follow similar steps. Here, however, we will take advantage of the Ethos-U55 to increase the performance of the system. Let us go over the software flow for the Ethos-U55, as this is relatively new to most developers.

Ethos-U55 in practice

Have a look at the following diagram:

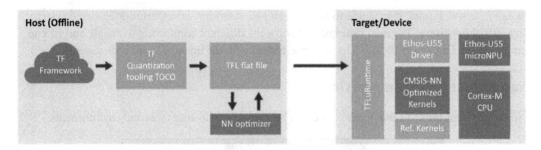

Figure 6.3 – Ethos-U55 optimized software flow

You can benefit from the processing capability of the Ethos-U55 microNPU by using the TFlite Micro ML framework from Google, which runs on the host application processor (Cortex-M55, in the case of the Corstone-300 FVP). The process starts by training or acquiring a TensorFlow model that is to be accelerated. The model is then quantized to 8-bit integer format and converted to the standard TensorFlow Lite flat-file format (`.tflite` file). The TensorFlow Lite flat file is then input to an NN optimizer tool called Vela, which you run on your host machine. Vela is an open sourced Python tool that you can install from the PyPI community (`https://pypi.org/project/ethos-u-vela/`). The output of the tool is an optimized TensorFlow Lite file that is now ready for deployment on your Ethos-U55 target device or, in this case, the Corstone-300 FVP.

Vela identifies which ML operators can be executed by the Ethos-U55 microNPU and substitutes these with a sequence of optimized and special operations. All other ML operators can be executed on the Cortex-M55 processor either by optimized kernels from the CMSIS-NN library or by fallback support in TensorFlow Lite reference kernels. The Ethos-U55 driver, which you can download from `https://review.mlplatform.org/plugins/gitiles/ml/ethos-u/ethos-u-core-driver`, manages the communication and workloads that execute the inferences on the Ethos-U55 microNPU.

Environment setup

The environment setup is identical to the previous section on anomaly detection; here are the commands again for easy reference.

Create and connect to an AVH AMI instance following these instructions: `https://arm-software.github.io/AVH/main/infrastructure/html/run_ami_local.html#connect`.

Then connect to the AMI over ssh (ensuring you have X forwarding enabled to view visual displays), and obtain the required resources with these commands:

```
ssh -i <your-private-key>.pem ubuntu@<your-ec2-IP-address>
git clone --recursive "https://review.mlplatform.org/ml/
ethos-u/ml-embedded-evaluation-kit"
cd ml-embedded-evaluation-kit/
python download_dependencies.py
python set_up_default_resources.py
```

At this point, you are set up to build the image classification application. You can also use the same AWS instance to run the anomaly detection and this image classification example; just keep the build directories separate to avoid overlapping files.

Build

The same cmake and Arm Compiler for Embedded tools will be used in this example as in the previous example. First, create a build directory, like so:

```
mkdir cmake-build-mps3-sse-300-ethos-u55-256-arm
cd cmake-build-mps3-sse-300-ethos-u55-256-arm
```

Then, build the example with the following two commands:

```
cmake .. -DTARGET_PLATFORM=mps3 -DTARGET_SUBSYSTEM=sse-300
-DCMAKE_TOOLCHAIN_FILE=~/ml-embedded-evaluation-kit/scripts/
cmake/toolchains/bare-metal-armclang.cmake -DETHOS_U_NPU_
ENABLED=ON -DETHOS_U_NPU_CONFIG_ID=Y256 -DUSE_CASE_BUILD=img_
class
make
```

This will again take about 2 minutes or so to complete, with the final executable being built as a result: ethos-u-image_class.axf. It is also located in the cmake-build-mps3-sse-300-ethos-u55-256-arm/bin/ directory. This time, because of the ETHOS_U_NPU_ENABLED=ON command, the Ethos-U55 will help run the application, set to use 256 MACs.

Run

To execute the image classification example, run the following command:

```
VHT_Corstone_SSE-300_Ethos-U55 -C ethosu.num_macs=256 -a ~/
ml-embedded-evaluation-kit/cmake-build-mps3-sse-300-ethos-u55-
256-arm/bin/ethos-u-img_class.axf
```

The Corstone-300 FVP simulation model can be run with several different parameter settings, which are passed with the -C option shown in the preceding code snippet. The ethosu.num_macs parameter maps to the number of 8x8 MACs performed per cycle. This should match the MAC setting we built the application for. By default, the simulation model sets this to 128. As we built our application for 256 MACs, we passed this value to the simulation model as well. As before, -a points to the application that we are running on the model.

With the application running, you will be presented with a menu of options, as shown here:

```
User input required
Enter option number from:

  1. Classify next ifm
  2. Classify ifm at chosen index
  3. Run classification on all ifm
  4. Show NN model info
  5. List ifm

Choice: 1
```

Figure 6.4 – Input selection for image classification example

The first menu option is to execute image classification on the next input image in the queue, which in this case is cat.bmp.

The following code snippet shows the application output for classification on the cat.bmp input image:

```
INFO - Running inference on image 0 => cat.bmp
INFO - Final results:
INFO - Total number of inferences: 1
INFO - 0) 282 (0.753906) -> tabby, tabby cat
INFO - 1) 286 (0.148438) -> Egyptian cat
INFO - 2) 283 (0.062500) -> tiger cat
INFO - 3) 458 (0.003906) -> bow tie, bow-tie, bowtie
INFO - 4) 288 (0.003906) -> lynx, catamount
INFO - Profile for Inference:
INFO - NPU IDLE: 784 cycles
INFO - NPU AXIO_RD_DATA_BEAT_RECEIVED: 2029001 beats
INFO - NPU AXIO_WR_DATA_BEAT_WRITTEN: 1151315 beats
INFO - NPU AXI1_RD_DATA_BEAT_RECEIVED: 432187 beats
INFO - NPU ACTIVE: 5426281 cycles
INFO - NPU TOTAL: 5427065 cycles
```

Figure 6.5 – Output from image classification example

The top five classifications with confidence indexes are provided, along with some performance data on how long it took the NPU to run an inference on the input cat image and provide the classification results. For example, in the preceding screenshot, you can see the cat.bmp image is classified as a tabby cat with a confidence of 75.39%.

You can similarly run an inference on any of the other input images in the example and check the classified output.

This section expanded on the previous anomaly detection example by running inferences on the Ethos-U55 through an image classification application. The next example, focusing on the third V, of voice, will dive deeper into another software stack to provide a wider perspective on ML software.

Micro speech – voice

To replicate the example in this section, you will be using the following tools and environments:

Platform	Arm Virtual Hardware – Corstone-300
Software	ML Application (Micro speech)
Environment	Amazon EC2 (AWS account required)
Host OS	Ubuntu Linux
Compiler	Arm Compiler for Embedded
IDE	-

Voice recognition is one of the most common ML example applications on Cortex-M microcontrollers. TensorFlow Lite is ideal for these use cases where a small memory footprint and optimized performance are important. This section demonstrates a TensorFlow Lite example application running on the same Cortex-M55-based AVH AWS AMI as the previous sections, now running a TensorFlow Lite example application called micro speech.

The micro-speech application is a simple demonstration of speech recognition. The application receives a Waveform audio file, also known as a wave or WAV file, as an audio sample. The micro-speech application is looking for the yes and no keywords. The Waveform file may contain the keywords the application is looking for or some other audio that is not one of the keywords. The input audio is processed by the application and reports if yes or no keywords are detected.

Micro speech is written in C++ and uses CMSIS build, also known as cbuild, to compile the application (different from the cmake build system used in the previous examples). CMSIS applications are constructed using software packs. One of the software packs used in micro speech is the TFlite Micro software pack. The application performance is also accelerated using CMSIS-NN kernels.

An additional explanation of the application source code, ML model, and Waveform file input processing will be covered in the following subsections.

Running ML models on microcontrollers

As we covered in the chapter introduction, ML applications consist of a training phase that takes test data and builds an ML model that is used during the inference phase to process new data. Model training is a complex subject beyond the scope of this text, but there are many good references to learn more about model training. To learn more about the training for the micro-speech application, look at the training details here: `https://github.com/tensorflow/tflite-micro/tree/main/tensorflow/lite/micro/examples/micro_speech/train`. To train an ML model for `yes` and `no` keyword recognition, a number of audio samples—including `yes`, `no`, and additional audio or background noise—are used. From this training data, the TensorFlow framework creates an ML model.

To use a model with TFlite Micro models must be saved in TensorFlow Lite format. The Lite format is an optimized format that can be processed with a very small amount of code and is easy to use in many languages, including C++. The file naming convention for ML models saved in the Lite format is to use a `.tflite` extension.

For the micro-speech application, we start from an already trained model that is saved for use as a TensorFlow Lite file (`.tflite`). The model is available on GitHub—the size is only 20 **kilobytes (KB)** and the amount of **random-access memory (RAM)** used is only 10 KB. Find the model here: `https://github.com/tensorflow/tflite-micro/blob/main/tensorflow/lite/micro/examples/micro_speech/micro_speech.tflite`.

Microcontrollers often run with no OS and may not have a filesystem to store ML models in `.tflite` format. Instead, the `xxd` command (`https://linux.die.net/man/1/xxd`) is used to turn the `.tflite` file into a **hexadecimal (hex)** data array that can be directly embedded in the C++ application. This means that changing the ML model will require the application to be recompiled, but for microcontrollers, this is an efficient solution.

To convert a `.tflite` file into a C++ file, run the following command:

```
xxd -i micro_speech.tflite > model.cc
unsigned char micro_speech_tflite[] = {
    0x20, 0x00, 0x00, 0x00, 0x54, 0x46, 0x4c, 0x33, 0x00, 0x00,
0x00, 0x00,
    0x00, 0x00, 0x12, 0x00, 0x1c, 0x00, 0x04, 0x00, 0x08, 0x00,
0x0c, 0x00,
<< continued >>
```

Add the `DATA_ALIGN_ATTRIBUTE` macro. This macro ensures that the model data array is aligned in a way that it does not overlap memory boundaries for optimal read performance by the processor.

This provides a basic overview of how to create an ML model for a microcontroller application.

Next, let's look at the tools used to make the micro-speech application.

Environment setup

The micro-speech example will be compiled using Arm Compiler for Embedded and run on AVH in an AMI, as with the previous examples in this chapter. The build process uses CMSIS packs (`https://github.com/ARM-software/CMSIS_5`) and the CMSIS-Toolbox (`https://github.com/Open-CMSIS-Pack/devtools`). The most important CMSIS pack for the project is the TensorFlow Lite pack (`https://github.com/MDK-Packs/tensorflow-pack`), which provides a build-and-test environment for TFlite Micro.

The micro-speech example is designed to be compiled and run on Linux. Additional tools that are part of the AMI are listed here:

- Python
- Ninja build system (`https://ninja-build.org/`)
- CMake
- Corstone-300 FVP
- Arm Compiler for Embedded
- The CMSIS-Toolbox, including `cbuild.sh` to build the application

To get started, connect to an AWS instance of the AMI using `ssh`, like so:

```
$ ssh -i <your-private-key>.pem ubuntu@<your-ec2-IP-address>
```

Once connected to the AMI, clone the project repository, as follows:

```
$ git clone https://github.com/ARM-software/AVH-TFLmicrospeech.git
```

Application overview

Before getting started, let's review the contents of the project to get familiar with the code.

The `Platform_FVP_Corstone_SSE-300_Ethos-U55/` directory contains all the files required to build and run the application on the FVP. Look in this directory for the project configuration and source code for the `main()` functions.

The project build is configured via a project file, `microspeech.Example.cprj`.

The `.crpj` file includes the following information:

- Required CMSIS packs
- Compiler to be used
- Compiler flags

- CMSIS components to be used
- Source files that make up the project

The ML model is located in the `./micro_speech/src/micro_features` subdirectory. Look at the `model.cc` and `model.h` files to see the ML model that was created using the flow described in the *Running ML models on microcontrollers* section earlier and the `xxd` command to generate an array of data found in `model.cc`.

The application starts from the `main()` function in `main.c` and is built around CMSIS-RTOS2 (`https://www.keil.com/pack/doc/CMSIS/RTOS2/html/index.html`), which provides generic **real-time OS (RTOS)** interfaces for Cortex-M microcontrollers. The `main()` function calls the `app_initialize()` function, which starts the main thread, named `app_main()`, in `microspeech.c`. This main thread calls the TensorFlow Lite `setup()` and `loop()` functions to begin the audio processing. The `loop()` function is called from an infinite loop to continuously process audio data.

The CMSIS-NN software library is also used. This provides a library of efficient NN kernels developed to get the best performance across a variety of Cortex-M microcontrollers while keeping memory usage low. For the Cortex-M55, it will take full advantage of the **Cortex-M Vector Extensions (MVE)** to provide high performance without the need to learn or use assembly code to implement vector operations.

The library covers a variety of function classes useful in ML applications, as follows:

- Convolution functions
- Activation functions
- Fully connected layer functions
- **Single-value decomposition filter (SVDF)** layer functions
- Pooling functions
- Softmax functions
- Basic math functions

We introduced this library of functions for operating on different weight and activation data types in *Chapter 2, Selecting the Right Software*. It may be interesting to browse the CMSIS-NN source code to get a feel for the type of functions available and how they are implemented. The micro speech uses 8-bit integer functions. The code can be found here: `https://github.com/ARM-software/CMSIS_5/tree/develop/CMSIS/NN/Source`.

The Corstone-300 FVP is used to execute the micro-speech application. Recall that the Corstone-300 is a full hardware design including the Cortex-M55 processor, the Ethos-U55 ML processor, memories, and peripherals.

Build

To build the application using `cbuild` from the CMSIS-Toolbox, run the following command:

```
$ cd Platform_FVP_Corstone_SSE-300_Ethos-U55
$ cbuild.sh microspeech.Example.cprj
```

It will take about a minute to compile the entire application and produce a binary at `Objects/microspeech.axf`. To run the application, execute the following script as shown in the command:

```
$ ./run_example.sh
```

This produces the output from the application, as shown here:

```
Fast Models [11.16.14 (Sep 29 2021)]
Copyright 2000-2021 ARM Limited.
All Rights Reserved.

telnetterminal0: Listening for serial connection on port 5000
telnetterminal1: Listening for serial connection on port 5001
telnetterminal2: Listening for serial connection on port 5002
telnetterminal5: Listening for serial connection on port 5003

Ethos-U rev afc78a99 --- Aug 31 2021 22:30:42
(C) COPYRIGHT 2019-2021 Arm Limited
ALL RIGHTS RESERVED

Heard yes (146) @1000ms
Heard no (145) @5600ms
Heard yes (143) @9100ms
Heard no (145) @13600ms
Heard yes (143) @17100ms
Heard no (145) @21600ms

Info: Simulation is stopping. Reason: Cycle limit has been
exceeded.

Info: /OSCI/SystemC: Simulation stopped by user.
[warning ][main@0][01 ns] Simulation stopped by user
```

```
--- cpu_core statistics: ------------------------------------
-----------------
  Simulated time                          : 23.999999s
  User time                               : 25.804117s
  System time                             : 3.336213s
  Wall time                               : 29.132544s
  Performance index                       : 0.82
  cpu_core.cpu0                           :   26.36 MIPS (
768000000 Inst)
  ----------------------------------------------------------
-----
```

The print statements from the software occur when the application has detected yes or no.

Let's look at where the input comes from and understand what it is and how to change it.

Run

The default audio input file is test.wav. You can download this file to your computer and play it using an audio player application. On Windows or macOS computers, the file should play automatically when you click on it. You should hear a sequence of yes, no three times. This corresponds to the printed output from the application.

The test.wav file is sent into the hardware design using Python. The Python interface to the simulated hardware is called the **Virtual Streaming Interface**, or **VSI** (https://arm-software. github.io/AVH/main/simulation/html/group__arm__vsi.html). VSI is a flexible, generic peripheral interface that can be used to simulate data streaming in and out of a device for applications such as audio, video, or sensor data. In the micro-speech application, VSI is used to stream the test.wav file into memory. This streaming interface takes the place of a microphone capturing audio. Streaming audio data from a file is useful for automated testing. It allows a large library of sample audio data to be saved and used over and over as the ML model is changed.

The Python code for the audio input is located in VSI/audio/python/arm_vsi0.py. Take a look at the code and see that it is reading test.wav when the audio driver running on the processor writes to the control register to initiate the streaming audio transfer. The audio driver is found in the VSI/audio/driver directory. The VSI is controlled by the audio driver.

Presented here are two recommendations for the next steps to try:

- Experiment with the micro-speech example by recording a new `test.wav` file and run it. Check whether the micro-speech application can correctly process your audio sample.

- Edit the `arm_vsi0.py` file to enable more logging and try to understand the sequence of events for how it works.

For the second option, uncomment the following line:

```
#verbosity = logging.DEBUG
```

Then, add a comment (#) in front of the following line:

```
verbosity = logging.ERROR
```

This changes the verbosity to DEBUG instead of ERROR, so more output will be printed when the application is run.

Use the `run_example.sh` script again and see the extra output coming from the Python code, and see whether you can understand how the audio streaming is working in conjunction with the audio driver to process the audio file.

This section has presented the TFlite Micro example called micro speech. It has explained numerous points that are different from the official version found in TFlite Micro We have reviewed the following points:

- Application creation with CMSIS packs and the CMSIS-Toolbox to build Cortex-M applications on Linux

- UsingAWS to start a **virtual machine** (**VM**) with the needed tools already installed to save time

- Running applications on an FVP instead of a physical board when the hardware is not available

- Providing streaming audio data from files instead of a microphone to create a library of test data that can be reused as software changes

Summary

In this chapter, we first provided an overview of the ML software development process. We also looked at ML frameworks and libraries that can be leveraged to build ML applications on Arm Cortex-M-based devices.

In the remainder of the chapter, we focused on the steps to run three different ML use cases on the Arm Cortex-M55 AVH simulation system.

In summary, we have looked at the ML software flow targeting Arm Cortex-M devices, components that enable this flow, and examples that illustrate running this flow. In the next chapter, we will cover a vaunted but often confusing area of embedded development: security. We will provide both an introduction to and an implementation guide for security in Cortex-M products.

Further reading

For more information, refer to the following resources:

- TinyML example:

  `https://dev.to/tkeyo/tinyml-machine-learning-on-esp32-with-micropython-38a6`

- Arm ML Embedded Evaluation kit:

  `https://review.mlplatform.org/plugins/gitiles/ml/ethos-u/ml-embedded-evaluation-kit/+/HEAD/docs/documentation.md#arm_ml-embedded-evaluation-kit`

- Blog post on the optimization of ML models running on Arm Ethos-U microNPU:

  `https://community.arm.com/arm-community-blogs/b/ai-and-ml-blog/posts/optimize-a-ml-model-for-inference-on-ethos-u-micronpu`

- TinyML audio classification example running on a Cortex-M0+ microcontroller board:

  `https://blog.tensorflow.org/2021/09/TinyML-Audio-for-everyone.html`

- TinyML motion recognition example on Raspberry Pi Pico:

  `https://mjrobot.org/2021/03/12/tinyml-motion-recognition-using-raspberry-pi-pico/`

- TensorFlow Lite for Microcontrollers: `https://www.tensorflow.org/lite/microcontrollers`

7

Enforcing Security

As the world grows more reliant on electronics, and especially with the addition of connectivity to IoT applications, security has become a vital concern. If a connected device stores any type of sensitive data – such as Wi-Fi passwords, certificates, or personal information – this data needs to be secured. There is a common sentiment that if a device is connected to the internet and has some value, someone will try and hack it.

Even if your hardware implements specific technology features for security such as TrustZone, if your software has security-related flaws, it can compromise your entire device. For modern developers, software must be architected with security as a primary driving requirement and not an afterthought. This chapter contains several examples of secure software implementation, but to truly build security into your device, it must be planned for from the start of a project.

To provide some guidance on how to holistically implement security into Cortex-M devices, in hardware and software, Arm introduced the **Platform Security Architecture (PSA)** framework. We will start by going over the four stages of PSA and then move on to running secure software examples.

In a nutshell, the following topics will be covered in this chapter:

- Breaking down PSA
- Example 1 – Secure versus non-secure hello world
- Example 2 – TF-M

Breaking down PSA

The PSA framework was introduced by Arm in 2017. It was created to reconcile the growing number of intelligent connected devices around the world with the fractured and varied approach to securing them. Cortex-M-based products are as numerous as they are different; any one device has hundreds of unique hardware or software products from dozens of companies inside it. If even one of these products has a security gap, the entire device has a security vulnerability. By 2017, the need was clear for an industry-wide framework to holistically address the security problem in a cost-effective manner. The PSA framework fills that need.

We briefly covered PSA in *Chapter 2, Selecting the Right Software*. There is a white paper available at `https://www.psacertified.org/blog/program-overview-digital-whitepaper/` with an in-depth explanation of the three parts of PSA, as listed here:

1. Threat models and security analyses, derived from a range of typical IoT use cases

2. Architecture specifications for firmware and hardware

3. Open source reference implementation of the firmware architecture specifications

The *threat models and security analyses* are available at `https://www.psacertified.org/development-resources/building-in-security/threat-models/`, and we will also dive deeper into the PSA's **10 security goals** in the upcoming section. The *architecture specifications for firmware and hardware* are available for free at `https://www.psacertified.org/development-resources/building-in-security/specifications-implementations/`.

The *open source reference implementation of the firmware architecture specifications* is the **Trusted Firmware-M** (**TF-M**) software stack, discussed at the end of this chapter with example implementations. In addition to the three parts of the PSA framework described previously, there is a formal program intended to verify that devices follow the PSA framework in practice: *PSA Certified*. This program defines three levels that chips, boards, and software stacks can be tested against to verify their robustness against security threats.

All these resources can be put together, enabling Cortex-M developers to select security-hardened hardware, develop resilient software, and certify that their product is built with the right level of security prioritized. While there is a lot of material to sort through, it may be helpful to think about each resource as contributing to one of these four stages of security design:

- **Analyze**: Understand the assets you are protecting and the potential security threats you face

- **Architect**: Create your design stemming from the identified security requirements

- **Implement**: Develop your product from secure hardware, firmware, and software stacks

- **Certify**: Obtain a PSA Certification through an independent third party

Analyze – threat modeling and the 10 security goals

The first step is to understand what use case your device is intended for and how it will be deployed through its life cycle. An IoT asset tracker is different from a smart door camera, for example.

Then, with your device's life cycle laid out, you need to understand the security threats that will exist in your device. Distinctions between use cases, such as being connected to the internet or physically accessible to the public, will affect the type of threats your device will face.

PSA provides a list of possible attack vectors and threat models to frame your analysis, which is fully accessible at `https://www.psacertified.org/app/uploads/2020/10/PSA_Certified_Threat_Model_and_Security_Goals_RC3.pdf`.

Here is a brief summary of the possible threats, organized into two categories:

- **Software attacks**

  - Spoofing the system identity, faking trustworthiness

  - System cloning to steal the device or app identity

  - Spoofing the firmware update sender's identity, injecting firmware

  - Image tampering, gaining system access

  - Update rollbacks for legacy bug exploitation

  - Software data tampering polluting trusted services

  - Persistent malware exploiting vulnerabilities across reset cycles

  - Deniability and erasing logs and history through various methods

  - Side-channel attacks extracting sensitive data

  - Software data extraction fooling communication interfaces

  - Abusing communication between trusted and non-trusted services

  - Unrestricted access through breached trusted services

- **Hardware attacks**:

  - Physical data tampering, reading or writing removable media

  - Physical data extraction, eavesdropping on physical bus lines

  - Physical debug abuses, viewing assets during repair or end-of-life

The list of possible attack vectors is always evolving and changing, but this list offers a good place to start when considering possible security vulnerabilities.

The final step in the analysis is to create security requirements based on these possible attack vectors. To guide you in this process, the Platform Security Model contains 10 security goals in the PSA framework. These goals are intended to provide an implementation-agnostic way to think about the essential features of a secured, trusted product. The 10 security goals provide a high-level common language to discuss security robustness rules, with specific implementations coming later in the process.

The full Platform Security Model document is very detailed and can be found at the same link provided for threat modeling on the PSA Certified website. Here we will briefly introduce each of the 10 security goals with a straightforward explanation:

- **Goal 1: Devices are uniquely identifiable**

  In order to be properly trusted, a device must be provably unique. For example, when connecting your phone to your new Bluetooth headphones, those headphones must be uniquely identifiable to ensure your phone connects to the intended target. If they are not unique, it becomes a security risk: you could accidentally connect to a malicious device posing as your headphones, or a man-in-the-middle attack could take place by intercepting communications before forwarding them along.

- **Goal 2: Devices support a security life cycle**

  Every product goes through many environments in its life, from creation to death. At each point in this life span, devices should be as secure as necessary to avoid threats. This includes taking proper precautions during development, ensuring a trustworthy manufacturing and deployment process, protecting the device virtually and physically when in use, and seeing it through a proper end-of-life. Different devices will need more attention at different stages. A connected power station management system should be physically protected from malicious actors, and, for example, a mass-produced smart lightbulb should have special protection against virus injection during the manufacturing stage.

- **Goal 3: Devices are securely attestable**

  The definition of attestable is as follows: *to affirm to be correct, true, or genuine*. This combines with the first two goals to assert that a device must be provably trustworthy at every point in its life cycle.

- **Goal 4: Devices ensure that only authorized software can be executed**

  This is a straightforward goal. In some cases, such as allowing user-inputted code, unauthorized software may have to be run on the device. In these cases, it should be ensured that this unauthorized software cannot compromise device security. Secure boot and secure loading processes are fundamental pillars in assuring this goal is met.

- **Goal 5: Devices support secure updates**

  For connected devices that are continually updated, ensuring security and authenticity during an update is essential. Multiple components of the device should be updatable, including security credentials, programmable hardware configurations, and software.

- **Goal 6: Devices prevent unauthorized rollbacks of updates**

  As devices are updated to fix security vulnerabilities, they should not be allowed to revert to a previous (and vulnerable) software version. This anti-rollback principle applies to both accidental and malicious attempts. Recovery of data is the only exception to this rule and should be handled with care.

- **Goal 7: Devices support isolation**

  Isolation contributes to hardening security by reducing attack surface areas. Even if an attacker breaks into one portion of the device (in hardware or software), the other portions are not accessible. These isolation boundaries apply between device software sections and the device software and the outside world.

- **Goal 8: Devices support interaction over isolation boundaries**

  Perhaps a clear requirement stemming from the previous goal is the need to interact over isolation boundaries to create a functional system. This interaction should not be able to compromise any individual section, however, and exchanged data must be validated as confidential and trustworthy.

- **Goal 9: Devices support the unique binding of sensitive data to a device**

  Sensitive data, such as secret keys or credentials, deserves to be treated with extra care. This goal mandates binding sensitive information to an individual device to prevent secret data from spreading elsewhere. It is also recommended to store sensitive data in inherently secure storage, enable access to select individual device users, and prevent data access in certain security states, such as during debugging.

- **Goal 10: Devices support a minimal set of trusted services and cryptographic operations that are necessary for the device to support the other goals**

  To minimize attack surfaces and enable security analysis, trusted software or services should be kept as small as possible. Only implement the security features required and implement them excellently.

You can use these security goals to address the identified attack vectors for your device and detail the list of specific security requirements for your project. At this point, you are ready to start architecting a solution to these security requirements.

Architect – References, suggestions, and APIs

In this step, you translate security requirements into an appropriate system in both hardware and software. The PSA framework does offer some documentation and examples to guide in this process, but naturally, some architecture decisions will have to be made regarding your specific project.

The guidance PSA provides in this architecture creation phase is broken into two areas: selecting existing hardware or software to use in your project and creating your own.

Selecting the right SoC, device, and software

There are detailed recommendations for creating a secure **system-on-chip** (**SoC**) for device makers to follow in the *Trusted Base System Architecture* document. It is too low-level to be useful in this context and all that you need to know is which SoCs have followed this security analysis process. This is where the concept of being *PSA Certified* comes in, which is covered as the last phase for your device to go through in this section. Just as end products can be certified as having factored in security decisions, so too can SoCs.

A guide to selecting the right SoC for your design can be found at `https://www.psacertified.org/blog/choosing-iot-security-chip/`. A key offering from Arm is the TrustZone technology, mentioned in *Chapter 1, Selecting the Right Hardware*, which creates physical isolation between critical firmware and the rest of the application. You will find SoCs that implement different levels of security best practices on top of Arm TrustZone technology in the Cortex-M space from that list.

Similarly, there is a list of software that has been independently certified to be fit for security applications. They can be found at `https://www.psacertified.org/certified-products/?_standard=psa-certified-system-software`. Lastly, if you are only looking to develop software on top of an existing hardware device, there is a list of certified devices available at `https://www.psacertified.org/certified-products/?_standard=psa-certified-oem` as well.

Creating secure hardware and software

The PSA Certified site contains varied resources for architecting security-focused systems. The full list of resources can be found at `https://www.psacertified.org/development-resources/building-in-security/specifications-implementations/`. It includes Arm-specific, low-level suggestions around hardware, firmware, and **Root of Trust** (**RoT**) requirements. Another resource is a list of functional APIs, easing the implementation of cryptography, secure storage management, and attestation functions.

Implement – Creating a secure system

This phase is about creating your device based on the proposed architecture that meets your identified attack vectors. Most of this book is focused on helping you implement software on Cortex-M devices, with examples being crucial for aiding and speeding up development. Similarly, the PSA offers an example implementation of PSA specifications on Cortex-M devices. TF-M is this reference and is open source for all to refer to and build from. We will dive into some TF-M examples later in this chapter.

Certify – PSA certification levels

The last step after creating a device with security in mind is to certify that your device meets the security requirements. This certification helps developers and customers trust and rely on this component to achieve the level of security they need. This is helpful for devices that will be used as a foundation to build other solutions from to prove that security was a key design consideration. We referenced the list of SoCs, devices, and software that are already certified in the *Architect – References, suggestions, and APIs* section (which can be found here: `https://www.psacertified.org/certified-products/`).

The certification process evaluates SoCs, devices, and software to be tested under laboratory conditions that measure their level of security. There are three progressive levels of this security certification:

- **Level 1: Security principles**

  For a device manufacturer, this is simply achieved by selecting pre-certified silicon and a pre-certified software platform for your device and then implementing the 10 aforementioned security goals. It requires the completion of a critical security questionnaire (based on the 10 security goals and identified threat models). Level 1 is also the limit for certification for devices, system software, and APIs. The latter levels are intended for SoCs building higher levels of security into the heart of devices.

> **Important note**
>
> Common software such as FreeRTOS from AWS and common SoC families such as ST STM32s and NXP i.MXs are all certified under Level 1.

- **Level 2: Software attacks**

  PSA Certified Level 2 is specifically for SoCs and provides evidence of protection against scalable software attacks. Chips are evaluated for 25 days at an independent laboratory, undergoing penetration testing and vulnerability analysis focusing on the **PSA Root of Trust (PSA-RoT)**. Chips must prove they can withstand the threats presented in the PSA-RoT Protection Profile, available here for those curious: `https://www.psacertified.org/getting-certified/silicon-vendor/overview/level-2/`. Chips at this level are best suited to protecting key assets that are available over software but do not offer physical access to attackers, such as various smart home devices inside the home.

- **Level 3: Hardware attacks**

  PSA Certified Level 3 is also specifically for SoCs and provides evidence of protection against more sophisticated attack types, including side channel and physical attacks. Chips are evaluated for 35 days and undergo rigorous security testing from software and physical attack vectors. It is best to select chips with this level of security for your device if protecting high-value assets that are publicly accessible, such as a smart home door lock. Information is available here: `https://www.psacertified.org/getting-certified/silicon-vendor/overview/level-3/`.

> **Important note**
>
> There is also a separate certification available for functional APIs that do not have independent levels. This is what the Keil RTX5 RTOS is certified under.

There is an extensive list of resources to obtain certification at any of these levels and those interested can view it on the PSA Certified website: `https://www.psacertified.org/development-resources/certification-resources/`. If you are interested in getting your Cortex-M device PSA Level 1 Certified, follow the steps previously outlined and submit your certification request through the PSA Certified website. More commonly, this understanding of PSA Certification is helpful to narrow down SoC purchasing decisions to those meeting your desired security requirements. This goes for selecting secure base firmware as well.

With this helpful PSA security framework in mind, we can narrow our focus back down to the implementation of secure software. In the remainder of this chapter, we will discuss how to start from security examples, learn how the software implements key security concepts in practice, and how to leverage the open source TF-M code through some example implementations on the Cortex-M33 and Cortex-M55.

Example 1 – Secure versus non-secure hello world

In general, when you start a new project for an Arm Cortex-M device with TrustZone (such as the Cortex-M33 and Cortex-M55), the project will comprise two sub-projects: a secure and non-secure project. Secure and non-secure code have their own boot code and are compiled and linked independently in the sub-projects. Both secure and non-secure code run on the same processor but are loaded in isolated and independent areas of memory. All the code that handles security and configuration, such as boot code, firmware updates, and crypto libraries, is placed in the secure project. The rest of the application code is placed in the non-secure project. The objective is to minimize the amount of code in the secure project and run exhaustive checks on it for security vulnerabilities.

> **Important note**
>
> The implementation of secure software with TrustZone inherently follows the 10 security goals – specifically, Goal 7 of isolating secure and non-secure areas; Goal 8 of enabling interaction between these areas; and Goal 10 of minimizing the number of secure services on a device. Thinking about implementing security in the framework of PSA's 10 security goals can help organize your thinking to better understand ostensibly complex security implementation.

Example 1 – Secure versus non-secure hello world   149

The secure versus non-secure code is managed and defined by the memory map settings. All physical memory addresses are assigned as either secure or non-secure. The assignment is managed by the **Security Attribution Unit (SAU)**, which is only accessible in the secure state. This provides independent memory protection for each security state. When an area of memory is addressed in software, the load and store instructions acquire the secure or non-secure attribute. Thus, non-secure access to a secure address will result automatically in a memory fault. A non-secure program can only call a secure function and return to the non-secure mode in a standardized way, explained in this upcoming example.

Here is a quick list of what the secure sub-project and non-secure sub-projects can and cannot do for reference:

- The secure sub-project can access everything

- The non-secure sub-project cannot access secure sub-project resources

- Both secure and non-secure code is running on the same processor and can implement independent scheduling

Now, with several security principles discussed, we will go through an example of them in context. This example walks through the secure version of *hello world* on the Cortex-M33-based NXP board, showing in detail how the processor switches from a secure to a non-secure state and how to call a secure function from a non-secure state.

To replicate the example in this section, you will be using the following:

Platform	NXP LPCXpresso55S69
Software	hello world (secure and non-secure)
Environment	Local PC
Host OS	Windows
Compiler	Arm Compiler for Embedded
IDE	Keil MDK

Obtaining and building the application

To start, open the Keil MDK IDE and get the examples through the **Pack Installer** widget. Select the **LPC55S69** device and copy **hello_world_ns** and **hello_world_s** into your workspace. If the **Copy** action is disabled, close all the projects open in your Keil MDK IDE and try again. In the window that pops up, check the **Use Pack Folder Structure** checkbox and uncheck the **Launch μVision** checkbox:

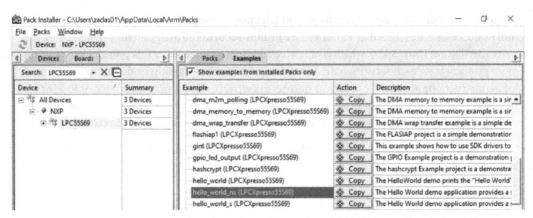

Figure 7.1 – The Pack Installer window with NXP LPC55S69 board examples

Once copied, in the IDE, select **Project | Open Project…** and navigate to the copied **hello_world_s** example. Under this example, you will find a μVision Multi-Project file called `hello_world.uvmpw`. If you copied the example to your `Documents` folder, you will find this file under this path:

Documents › trustzone_examples › hello_world › cm33_core0 › hello_world_s › mdk

Figure 7.2 – File navigation to the hello world Trustzone example

Upon opening this μVision Multi-Project File, both the sub-projects will be loaded in the same view. Keil MDK tools provide something called a *multi-project workspace view* for compiling and running TrustZone-based applications that have both a secure and non-secure sub-project associated with it. Other IDEs such as MCUXpresso also provide a similar feature to make the development of these applications easier. This way, you can control and view the source files for both applications that are going to run on the same target without switching between project spaces. The following figure shows what this looks like on your screen:

Example 1 – Secure versus non-secure hello world   151

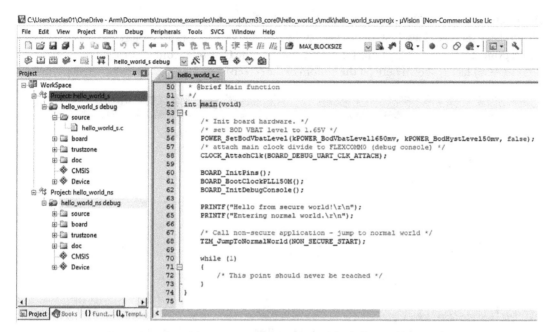

Figure 7.3 – A multi-project workspace view of the hello world example

By default, this project is configured to use the Arm Compiler for Embedded toolchain to compile the application. Arm provides a feature called **Cortex-M Security Extension** (**CMSE**) in the compiler for the development of secure code on Cortex-M processors. It is documented in detail at https:// developer.arm.com/documentation/100720/0200/CMSE-support. CMSE features are available in other compiler toolchains such as GCC as well. An important C pre-processor macro, __ARM_FEATURE_CMSE, is one of the features that indicates whether code is running in secure or non-secure mode. The access to secure and non-secure aliases for all peripherals is managed using this compiler macro.

There are two function attributes to support calls between secure and non-secure modes:

- __attribute__((cmse_nonsecure_entry)): A secure function that can be called by non-secure code

- __attribute__((cmse_nonsecure_call)): A call to a non-secure function from secure code

To enable the CMSE support attributes and the usage of the pre-processor macros, you will need to compile your software project by passing the following options to the Arm Compiler for Embedded:

```
-mcpu=cortex-m33 -mcmse
```

This is set in the Keil µVision example by selecting the **Secure Mode** software model on the **Target** tab. To check these settings, first, right-click on the `hello_world_s` sub-project and select **Set as Active project**. Then, select **Options for Target** and view the **Compiler control string** setting on the **C/C++ (AC6)** tab, as shown here (at the bottom of the screen):

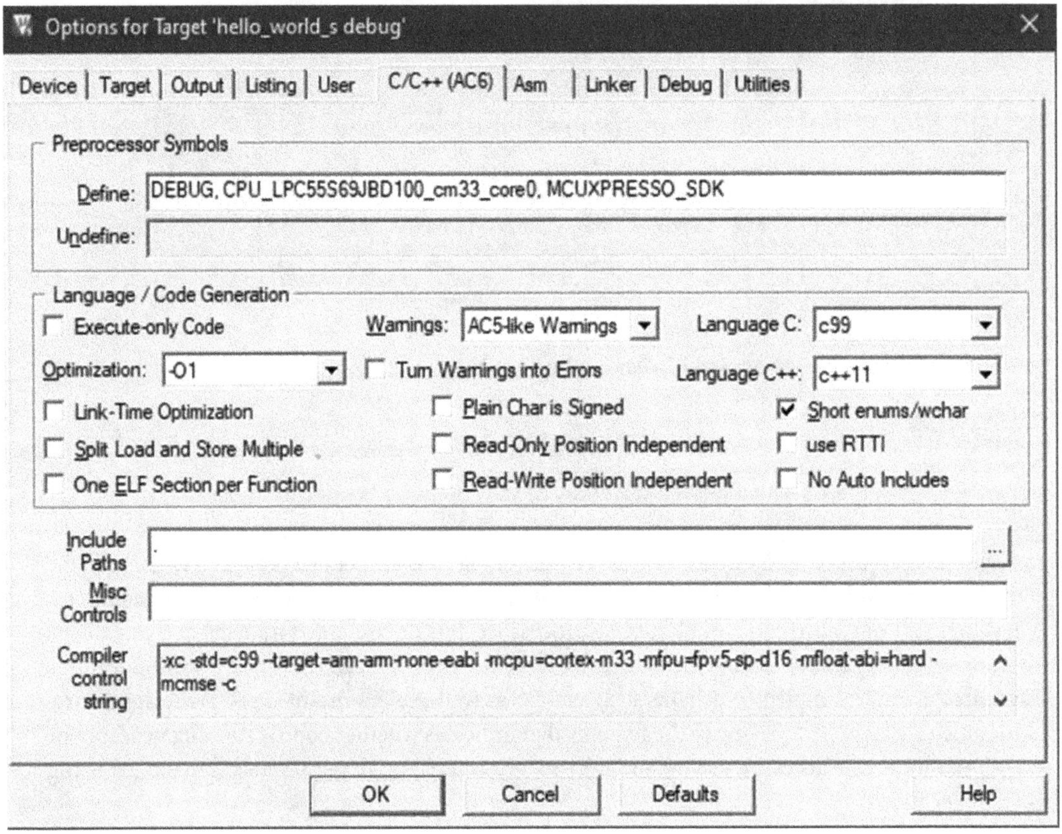

Figure 7.4 – A screenshot of the compiler options

To understand the example better, enable the creation of some helpful debug files before building. Again, select the **Options for Target** button, go to the **Output** tab, and check the **Debug Information** and **Browse Information** checkboxes.

Example 1 – Secure versus non-secure hello world   153

We will also generate the disassembly file for this application so that we can view and highlight the important instructions used in the context of secure and non-secure programming. To generate the disassembly file, select **Options for Target**, go to the **User** tab, and enter the following command to run after the build or re-build of the application is done. Copy it under `After Build/Re-build > Run #1` and make sure the **Run #1** checkbox is checked:

```
fromelf $Pdebug\%L --disassemble --interleave=source  --text -c
--output=$Pdebug\hello_trustzone.dis
```

Now, build the project by selecting the **Batch Build** button (this builds both sub-projects at once).

Once the program is built, select the **Start/Stop Debug Session** option from the **Debug** menu to bring up the debug view; this will automatically load the contents of both `hello_world_s` and `hello_world_ns` onto the board. You can also use the *Ctrl + F5* keyboard hotkey to accomplish the same thing. Note that the `hello_trustzone.dis` file is also automatically created as part of the build, which contains the disassembly contents.

Switching security states

You should now be actively debugging the hello world example on the Cortex-M33 NXP board. At the start of debugging, the program counter (also known as the `r15` register) will be at the start of `main()` in `hello_world_s.c`. This means that the secure mode copy of `startup_LPC55S69_cm33_core0.s` has already been executed at this point. The TrustZone configuration code is always contained in the secure sub-project and executed right after reset. This is performed by the `BOARD_InitTrustZone()` function called from `SystemInit()` during startup.

At the start of `main()`, the board hardware is initialized and there are a couple of `printf` statements, as shown here:

```
    PRINTF("Hello from secure world!\r\n");
    PRINTF("Entering normal world.\r\n");
```

Up until this point in the application, the processor is running in secure mode. After these `printf` statements, a function called `TZM_JumpToNormalWorld(NON_SECURE_START)` is executed. This function initiates the processor's jump to the non-secure world.

We will now view the contents of this function in our disassembly file, `hello_trustzone.dis`, to understand what is going on here in more detail. Also, note that your disassembly file may look different if you have changed the default optimization settings or compiler version:

```
TZM_JumpToNormalWorld
;;; .\tzm_api.c (351)
        0x10000b48:    b580        ..      PUSH    {r7,lr}
        0x10000b4a:    f64e5208    N..R    MOV     r2,#0xed08
        0x10000b4e:    f2ce0202    ....    MOVT    r2,#0xe002
        0x10000b52:    6801        .h      LDR     r1,[r0,#0]
        0x10000b54:    f3818888    ....    MSR     MSP_NS,r1
        0x10000b58:    6010        .`      STR     r0,[r2,#0]
        0x10000b5a:    6840        @h      LDR     r0,[r0,#4]
        0x10000b5c:    e92d0ff0    -...    PUSH    {r4-r11}
        0x10000b60:    f0200001    ...     BIC     r0,r0,#1
        0x10000b64:    b0a2        ..      SUB     sp,sp,#0x88
        0x10000b66:    ec2d0a00    -...    VLSTM   sp
        0x10000b6a:    4601        .F      MOV     r1,r0
        0x10000b6c:    4602        .F      MOV     r2,r0
        0x10000b6e:    4603        .F      MOV     r3,r0
        0x10000b70:    4604        .F      MOV     r4,r0
        0x10000b72:    4605        .F      MOV     r5,r0
        0x10000b74:    4606        .F      MOV     r6,r0
        0x10000b76:    4607        .F      MOV     r7,r0
        0x10000b78:    4680        .F      MOV     r8,r0
        0x10000b7a:    4681        .F      MOV     r9,r0
        0x10000b7c:    4682        .F      MOV     r10,r0
;;; .\tzm_api.c (357)
        0x10000b7e:    4683        .F      MOV     r11,r0
        0x10000b80:    4684        .F      MOV     r12,r0
        0x10000b82:    f3808c00    ....    MSR     APSR_nzcvqg,r0 ;
formerly CPSR_fs
        0x10000b86:    4784        .G      BLXNS   r0
        0x10000b88:    ec3d0a00    =...    VLLDM   sp
        0x10000b8c:    b022        ".      ADD     sp,sp,#0x88
        0x10000b8e:    e8bd0ff0    ....    POP     {r4-r11}
        0x10000b92:    bd80        ..      POP     {r7,pc}
```

Figure 7.5 – The disassembly contents of TZM_JumptoNormalWorld

The TZM_JumpToNormalWorld(NON_SECURE_START) function – which is defined in the tzm_api.c source file – sets up the non-secure main stack and vector table first. It then gets a pointer to the non-secure resethandler and calls it. The non-secure resethandler function pointer type has __attribute__((cmse_nonsecure_call)). This attribute directs the compiler to generate a BLXNS instruction.

The execution of the BLXNS instruction here at the 0x10000b86 address is important to understand. Normally, you would expect to see a BX LR instruction to return to the calling code, but here, we have the BLXNS r0 instruction, which makes the processor switch from a secure to a non-secure state and branch to the address specified in the r0 register. This instruction was introduced as part of the Armv8-M architecture to enable security state switching. All the MOV rx or r0 instructions before the execution of this BLXNS instruction clear the register's contents to prevent any possible data leakage from the secure state.

Example 1 – Secure versus non-secure hello world   155

Calling a secure function from a non-secure state

At this point of execution, the processor is in non-secure mode. In non-secure mode, the processor will not be able to access the memory and peripherals of the secure world. The Cortex-M33 processor now executes the non-secure copy of the `startup_LPC55S69_cm33_core0.s` file and gets to the `main()` function in the non-secure hello world source file, `hello_world_ns.c`.

In this `main()` function, you will see two `printf` statements first:

```
PRINTF_NSE("Welcome in normal world!\r\n");
PRINTF_NSE("This is a text printed from normal world!\
r\n");
```

The `PRINTF_NSE()` function – which is defined in `veneer_table.c` – has `__attribute__` `((cmse_nonsecure_entry))` defined and uses it to call the `PRINTF` function, which is declared in the secure code. The function was created at secure entry and the attribute is leveraged to call it properly from non-secure mode.

The `PRINTF_NSE()` function – which is defined in `veneer_table.c` – has `__attribute__` `((cmse_nonsecure_entry))`. This attribute is used to call the `PRINTF` function, which is declared in the secure code scope. This is an example of using the `cmse_nonsecure_entry` attribute.

Next, the `StringCompare_NSE` function is called, which also has the same attribute and is defined in `veneer_table.c` as well. The function compares two strings by using a non-secure callback. The processor continues to be in a non-secure state when this function is called:

```
result = StringCompare_NSE(&strcmp, "Test1\r\n", "Test2\r\n");
    if (result == 0)
    {
        PRINTF_NSE("Both strings are equal!\r\n");
    }
    else
    {
        PRINTF_NSE("Both strings are not equal!\r\n");
    }
```

This example highlights how functions can be called from within the non-secure state, isolated from secure resources.

To investigate the exact instructions being executed, we can inspect the disassembled instructions for these functions from the `hello_trustzone.dis` file:

```
DbgConsole_Printf_NSE
        0x1000fe00:    e97fe97f    ....    SG          ; [0x1000fc08]
        0x1000fe04:    f7f0bd42    ..B.    B
__acle_se_DbgConsole_Printf_NSE ; 0x1000088c
    StringCompare_NSE
        0x1000fe08:    e97fe97f    ....    SG          ; [0x1000fc10]
        0x1000fe0c:    f7f0bdc2    ....    B
__acle_se_StringCompare_NSE ; 0x10000994
```

Figure 7.6 – The disassembly contents of the Printf_NSE function

Both functions execute an SG instruction. This is a Secure Gateway instruction. As defined in the Armv8-M instruction architecture, non-secure software can only call a secure function if the first instruction is a SG instruction and is in a memory region marked as non-secure callable (as shown in this code example). If these conditions are not met, then it results in a security violation. Because of these strict requirements, non-secure code cannot just jump to the middle of a secure function and run, providing further security isolation.

At the execution of the SG instruction, the processor switches from the non-secure state to a secure state. If you scroll through the disassembly towards the end of execution of the StringCompare_NSE function, you will see a BXNS lr instruction:

```
        0x10000abc:    4774        tG      BXNS        lr
        0x10000abe:    bf00        ..      NOP
```

On the execution of this instruction, the processor returns to non-secure mode. The BXNS, BLXNS, and SG instructions are all only available in a secure mode.

Having broken down the key parts of this application, we can let this program run to completion and view the actual program output. If you have not already, run the program by selecting the **Run** button in μVision. Make sure you are connected to the UART port to view the output coming through; I (Pareena) open the PuTTY tool and connect to UART over COM4 with a speed of 115200, as shown in the following screenshot. For a refresher on connecting to view output, review *Chapter 4, Booting to Main*:

Example 1 – Secure versus non-secure hello world   157

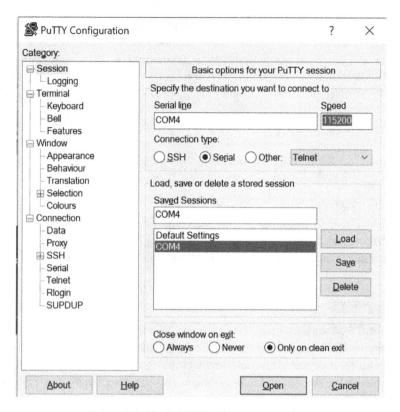

Figure 7.7 – The PuTTY Configuration window

You should see the following output:

```
Hello from secure world!
Entering normal world.
Welcome in normal world!
This is a text printed from normal world!
Comparing two string as a callback to normal world
String 1: Test1
String 2: Test2
Both strings are not equal!
```

As indicated by the printed messages, the overall flow of this application is to boot in the secure world, then switch to the non-secure world to call some secure functions, and finally, exit.

Through this example, we saw in detail how independent secure and non-secure software in the same project interact with and are isolated from one another. Understanding this interaction will help you to implement secure applications based on the PSA's 10 security goals. The most confusing area of secure software development is often this interaction of secure versus non-secure services and knowing the basic principles involved places you in a good position to develop your own secure applications.

The next example will address how to implement more of the security goals through the PSA reference implementation, TF-M.

Example 2 – TF-M

As noted previously, TF-M is a reference implementation of PSA for Cortex-M-based platforms. TF-M implements PSA developer APIs and has initially been targeted to Armv8-M architecture cores. It is reliant on the isolation boundary between the **Secure Processing Environment** (**SPE**) and **Non-Secure Processing Environment** (**NSPE**) that we covered in the previous example. It can be broadly broken down into three components.

Let's review each component in detail:

- **Secure boot**: TF-M software needs a secure bootloader that authenticates the integrity of the runtime images. This helps achieve Security Goal 4 regarding secure boot. TF-M currently uses a two-stage secure bootloader that validates that the images are from a trustworthy source and only then passes the right of execution to them. This implies all images in TF-M should be hashed and digitally signed for authentication purposes.

  TF-M uses MCUBoot as the secure bootloader. MCUBoot is open source and available on GitHub here: `https://github.com/mcu-tools/mcuboot`. It is automatically downloaded by the TF-M framework during the build process. The public signing keys are built into the MCUBoot bootloader and it allows for separate keys for signing secure and non-secure images.

  The following flow describes the secure boot execution by the processor at a high level:

  - At reset, the secure boot loader is started. As the name suggests, it runs in a secure mode. It runs the authentication process by verifying the digital signatures in the runtime images.

  - On successful validation, it passes control to the secure image, which initializes the environment, including the SAU, the MPU, and secure services. Only after that does it passes control to the non-secure image for execution.

- **Secure core**: The TF-M core features comprise secure system initialization and secure API calls invoked from both the SPE and NSPE and handled through secure IPC. It also features the **Secure Partition Manager** (**SPM**), which creates a database of secure partitions and sets up an isolation boundary between these partitions. A secure partition is a single thread of execution and the smallest unit of isolation. The principles regarding secure versus non-secure software isolation and interaction were discussed in the previous example.

Example 2 – TF-M   159

- **Secure services**: There are several secure services offered through TF-M. The **TF-M Cryptographic Secure (TF-M Crypto)** service provides a PSA Crypto API implementation in a PSA-RoT secure partition. PSA Secure Storage services provide a key-value storage interface for accessing device-protected storage. Essentially, the TF-M Crypto service allows your software to access common cryptography functions such as hashing and authenticated encryption.

  There are two ways to securely store information through TF-M. The PSA **Internal Trusted Storage (ITS)** is intended to store a small amount of security-critical data, such as cryptographic keys and firmware image hashes. This type of storage is offered by the PSA-RoT as a service and is accessible from the SPE side only.

In contrast, the PSA **Protected Storage (PS)** is intended to store larger data sets that are stored securely in an external flash, protected against physical attacks. This type of storage is offered at the application-level RoT as a service and is accessible from the NSPE as well as the SPE. It offers data-at-rest protection and can be configured to include device-bound encryption, integrity, or rollback protection.

> **Important note**
>
> More information about these storage types can be found in the PSA Storage API document, currently at version 1.0 at the time of writing: `https://armkeil.blob.core.windows.net/developer/Files/pdf/PlatformSecurityArchitecture/Implement/IHI0087-PSA_Storage_API-1.0.0.pdf`.

Another secure service offered in TF-M is the **TF-M Initial Attestation Service**. This service enables a device to prove its identity when needed. It can create tokens on request, containing a fixed set of device-specific items that are used to verify the device's integrity and trustworthiness by an external verifier.

> **Important note**
>
> There are other services offered by TF-M; a full list for your reference can be found in the TF-M documentation here: `https://tf-m-user-guide.trustedfirmware.org/docs/integration_guide/services/index.html`.

Now that we have a better understanding of the three TF-M components, let's walk through the process of building and running TF-M tests on the Corstone-300 FVP running in the AVH AMI.

Obtaining and building the application

The TF-M code described in this section focuses on testing the functionalities of various TF-M components. This includes the secure core component and assorted secure partitions. We can use this test suite to quickly get started with the TF-M code base and subsequently investigate its offerings.

To replicate the example in this section, you will be using the following:

Platform	Corstone-300 AVH
Software	Trusted Firmware-M
Environment	AWS EC2
Host OS	Ubuntu
Compiler	Arm Compiler for Embedded
IDE	-

Start by launching the AVH AMI and then SSH into the EC2 instance as you have with the examples described in previous chapters. To refresh your memory on this process, you can look at *Chapter 4, Booting to Main*, in the *Arm Virtual Hardware using the Cortex-M55* section, which describes this startup process in detail.

Now, on the EC2 instance running the AVH AMI, follow these commands to update the packages and install some Python prerequisites to build the TF-M suite of tests:

```
sudo apt update
sudo apt install python3.8-venv
sudo ln -s /usr/local/bin/pip3 /usr/bin/pip3.8
python3.8 -m pip install imgtool cbor2
python3.9 -m pip install imgtool cffi intelhex cbor2 cbor
pytest click
```

Next, we clone the TF-M repository and configure cmake to build the TF-M tests:

```
git clone https://git.trustedfirmware.org/TF-M/trusted-
firmware-m.git
cd trusted-firmware-m
mkdir cmake_build
cd cmake_build
cmake .. -DTFM_PLATFORM=arm/mps3/an552 -DTEST_NS=ON
```

You can either build all the TF-M tests or select a subset of them to build. The preceding commands include the -DTEST_NS=ON flag, which specifies only building the non-secure suite of tests. There are several different options you can pass to this cmake command to customize the suite of TF-M tests that you would like to run on the Corstone-300 FVP simulation target. They are all documented here: https://tf-m-user-guide.trustedfirmware.org/technical_references/instructions/tfm_build_instruction.html.

Example 2 – TF-M   161

As part of the build process, the TF-M tests repo is pulled from `https://git.trustedfirmware.org/TF-M/tf-m-tests.git` and the configured suite of tests is built. The TF-M test framework consists of both a secure and non-secure test suite, and for now, we are only using the non-secure side. The TF-M tests primarily target the functionality of the TF-M core and secure services API. Some of the checks performed by the TF-M test suite include the following:

- Validating the **inter-process communication** (**IPC**) interface between the isolated secure and non-secure firmware partitions as specified by the PSA framework

- Testing the two PSA Storage APIs: PS and ITS APIs

- Testing of the tokens created by the initial attestation service to verify the device identity during the authentication process.

- Validating the TF-M Crypto service, which allows the application to use ciphers, authenticated encryption, or hashes

The best way to look at what each of these tests is doing is by walking through the source code itself. For the IPC non-secure interface test we have just built, you can refer to the source code here: `https://git.trustedfirmware.org/TF-M/tf-m-tests.git/tree/test/secure_fw/suites/spm/ipc/non_secure/ipc_ns_interface_testsuite.c`. This suite has tests to connect to a secure RoT service from a NSPE or call the secure RoT service over an established connection.

You can inspect other tests in the repository to decide which ones you want to build and test in the same way, or just build them all. After you configure `cmake` to build the tests you are interested in executing, run the following command to build them:

```
make install
```

Upon a successful build, the TF-M test binaries will be created in the `bin/` directory. This includes binaries files for the MCUBoot bootloader, TF-M secure firmware, and TF-M non-secure application. Signed variants of both the TF-M secure and non-secure images are created, along with a combined signed image of both the secure and non-secure images.

In addition to the generated binaries, you can run the `fromelf` command to generate the disassembly files for both the secure and non-secure images:

```
fromelf -c tfm_s.axf > tfm_s.dis
fromelf -c tfm_ns.axf > tfm_ns.dis
```

We will use the disassembly files later in the example to inspect the instructions generated by the compiler for some of the test function calls.

Running the test suite

Now that we have successfully built the TF-M non-secure suite of tests, we are ready to run it on the Corstone-300 FVP that is already installed on the AMI.

Use this command to launch the simulation:

```
VHT_Corstone_SSE-300_Ethos-U55 -a cpu0*="bin/bl2.axf" --data
"bin/tfm_s_ns_signed.bin"@0x01000000
```

The `bl2.axf` file is the MCUBoot bootloader image that runs on the Cortex-M55 in the Corstone-300 FVP. `tfm_s_ns_signed.bin` is the combined signed image for the TF-M secure and non-secure image, and the address after the @ sign indicates where in the Corstone-300 FVP memory the image is loaded.

For your reference, the memory map for the FVP is documented here: `https://developer.arm.com/documentation/100966/latest/Arm--Corstone-SSE-300-FVP/Memory-map-overview-for-Corstone-SSE-300`.

The memory in the FVP is aliased for secure and non-secure usage. This is per the memory aliasing concept that was introduced with the Armv8-M architecture. To determine whether it is a secure or non-secure partition, bit 28 of the address is used by the memory protection controller on the system. What this implies is that the contents of `0x01000000` are the same as `0x11000000`, with the former being secure and the latter non-secure.

On running the preceding command, you should see a successful simulation run, which will output the `PASSED` status on each individual test. A summary similar to the one shown here is printed at the end of the simulation:

```
*** Non-secure test suites summary ***
Test suite 'IPC non-secure interface test (TFM_NS_IPC_
TEST_1XXX)' has PASSED
Test suite 'PSA protected storage NS interface tests (TFM_NS_
PS_TEST_1XXX)' has PASSED
Test suite 'PSA internal trusted storage NS interface tests
(TFM_NS_ITS_TEST_1XXX)' has PASSED
Test suite 'Crypto non-secure interface test (TFM_NS_CRYPTO_
TEST_1XXX)' has PASSED
Test suite 'Platform Service Non-Secure interface tests(TFM_NS_
PLATFORM_TEST_1XXX)' has PASSED
Test suite 'Initial Attestation Service non-secure interface
tests(TFM_NS_ATTEST_TEST_1XXX)' has PASSED
Test suite 'QCBOR regression test(TFM_NS_QCBOR_TEST_1XXX)' has
PASSED
```

Example 2 – TF-M   163

```
Test suite 'T_COSE regression test(TFM_NS_T_COSE_TEST_1XXX)'
has PASSED
Test suite 'TFM IRQ Test (TFM_IRQ_TEST_1xxx)' has PASSED


*** End of Non-secure test suites ***
```

Analyzing a RoT service connection test

Let us inspect one of the tests we have run to understand the program flow a little better. We will look at the `tfm_ipc_test_1003` test in `https://git.trustedfirmware.org/TF-M/tf-m-tests.git/tree/test/secure_fw/suites/spm/ipc/non_secure/ipc_ns_interface_testsuite.c`.

This test checks the connection of the NSPE to a RoT Service using its **secure function ID (SID)**. This is the test source code:

```
static void tfm_ipc_test_1003(struct test_result_t *ret)
{
    psa_handle_t handle;

    handle = psa_connect(IPC_SERVICE_TEST_BASIC_SID,
                         IPC_SERVICE_TEST_BASIC_VERSION);
    if (handle > 0) {
        TEST_LOG("Connect success!\r\n");
    } else {
        TEST_FAIL("The RoT Service has refused the connection!\
r\n");
        return;
    }
    psa_close(handle);
    ret->val = TEST_PASSED;
}
```

Both the NSPE and SPE use the following APIs to call the RoT secure services:

- `psa_connect`: This API is used to connect to a RoT secure service by its SID

- `psa_call`: This API calls a RoT secure service on an established connection

- `psa_close`: This API is used to a close a connection to the RoT secure service

The psc_connect() API calls the tfm_ns_interface_dispatch() function, which takes the SID and version, as shown in the function here:

```
psa_handle_t psa_connect(uint32_t sid, uint32_t version)
{
    return tfm_ns_interface_dispatch(
                                (veneer_fn)tfm_psa_connect_
veneer,
                                sid,
                                version,
                                0,
                                0);
}
```

The tfm_ns_interface_dispatch() function then creates a lock around this critical section of code and the API call structure is sent to the secure veneer function, tfm_psa_connect_veneer.

Inspect the tfm_s.dis disassembly file we generated for the secure image. You will see that this veneer function uses a SG instruction to enter secure mode before branching to the actual function, __acle_se_tfm_psa_connect_veneer, in the secure partition code:

```
tfm_psa_connect_veneer
    0x11000678:     e97fe97f     ....    SG       ; [0x11000480]
    0x1100067c:     f008beda     ....    B.W
__acle_se_tfm_psa_connect_veneer ; 0x11009434
```

Figure 7.8 – The disassembly contents of tfm_psa_connect_veneer function

The principle is the same as we saw in the previous section with the simple hello world TrustZone example. When non-secure code calls a secure function, it executes a SG instruction in a special veneer region first and only then branches to the secure function code.

Now, the API call is sent to the SPM. A handle for the connection is returned – the mutex around this critical code is also released:

```
__acle_se_tfm_psa_connect_veneer
    0x11009434:     ed6dcf81     m...    STCL     p15,c12,[sp,#-0x204]!
    0x11009438:     b500         ..      PUSH     {lr}
    0x1100943a:     f016fe18     ....    BL       psa_connect_cross ;
0x1102006e
    0x1100943e:     f85deb04     ]...    POP      {lr}
    0x11009442:     ec9f0a10     ....    VLDM     pc,{s0-s15} ; ?   ;
[0x11009484] = 0
    0x11009446:     e89f900e     ....    LDM      pc,{r1-r3,r12,pc} ; ?
    0x1100944a:     ecfdcf81     ....    LDCL     p15,c12,[sp],#0x204
    0x1100944e:     4774         tG      BXNS     lr
```

Figure 7.9 – The disassembly of tfm_psa_connect_veneer and return to non-secure

At the end of the execution of this secure function, a `BXNS  lr` instruction is executed, as shown in the disassembly. This causes a branch to the link register and a transition from the secure function code back to the non-secure code.

A similar program flow is followed to then close the connection to the RoT by calling the `psa_close` function. Upon a successful connection opening and closing, the test passes.

This TF-M example suite of tests, and the TF-M code base in general, is a powerful resource for understanding how to implement secure software on Cortex-M processors. You can use it as the foundation for your secure device or simply as an example reference to learn from. To leverage TF-M right away for your device, check to see whether your hardware is on the supported platform list: `https://tf-m-user-guide.trustedfirmware.org/platform/index.html`. If it is not, there are detailed instructions on how to add a new platform for TF-M support located here: `https://tf-m-user-guide.trustedfirmware.org/integration_guide/platform/index.html`.

Basing your software on TF-M and selecting a PSA Certified SoC or device are excellent ways to meet the PSA 10 security goals and build security directly into your device from the start.

Summary

This chapter outlined the security landscape for Cortex-M-based systems. We looked at the PSA framework first, which offers guidelines on how to systematically build security into your device. The first step is to analyze threats to understand the level of security needed for your specific use case. The second step is to architect a solution to plan what security needs to be implemented and how. The third step is to implement, build, or integrate your defined solution. The fourth step is to optionally certify your device's security.

We then implemented a secure versus non-secure state interaction through a hello world example on a Cortex-M33, breaking down how the two states manage interaction securely. Finally, we implemented a TF-M software test suite on a Cortex-M55, analyzing more security implementations in a realistic context.

Implementing proper security on Cortex-M devices can be a tricky undertaking. With the skills learned and resources available in this chapter, however, you are well positioned to ensure your next Cortex-M-based device is properly secured.

These past four chapters (*Chapter 4, Booting to Main*; *Chapter 5, Optimizing Performance*; *Chapter 6, Leveraging Machine Learning*; and *Chapter 7, Enforcing Security*) have all focused on delivering quality software by offering good coding practices, helpful libraries, and development frameworks. The next two chapters will focus on delivering quality software through a different means: useful development techniques. Up next, we discuss how to utilize the tools and services available through popular cloud providers to streamline your embedded development.

Further reading

To learn more about the topics that were covered in this chapter, take a look at the following resources:

- *ARMv8-M Security Extensions: Requirements on Development Tools – Engineering Specification*: `https://developer.arm.com/documentation/ecm0359818/latest/`

- *Secure software guidelines for Armv8-M*: `https://developer.arm.com/documentation/100720/0300`

- TF-M documentation: `https://tf-m-user-guide.trustedfirmware.org/`

- PSA Certified 10 goals explained: `https://www.psacertified.org/blog/psa-certified-10-security-goals-explained/`

- A guide on the TrustZone technology for Armv8-M architecture: `https://developer.arm.com/documentation/100690/latest/`

8
Streamlining with the Cloud

Software development for Cortex-M microcontrollers is steadily moving to the cloud. In this chapter, we will look at several current topics related to software development using the cloud. Historically, embedded software has lagged behind general-purpose software development when it comes to the adoption of new tools and methodologies. One cause is the conservative nature of embedded developers, which is understandable given the long life span of products requiring some stability. Another reason is that embedded software runs on microcontrollers that are not available in the cloud. Unlike general-purpose software, microcontroller software cannot run on general-purpose computers. Even with these limitations, the benefits of cloud development are becoming compelling for embedded Cortex-M projects.

The cloud is a common slang term referring to remote servers available over the internet and located in data centers around the world. With cloud computing, users don't purchase and manage physical servers. Instead, they rent servers using a variety of business models based on the *pay-for-what-you-use* concept. There are multiple cloud providers and hundreds of different services available for software developers to use during the creation and testing of software, as well as in the deployment of software products. In this chapter, we will focus on the embedded software developer and the services that contribute to the improved efficiency of development.

AWS was the early pioneer in cloud services, but similar features are available from other cloud providers such as Microsoft, Google, and Oracle. Much of what we will cover can be translated to other cloud providers, but we use AWS most of the time. This chapter will cover how to use virtual machines, the fundamental unit of cloud computing, as remote computers for development. Development containers are also an important technology used for remote and local development. We will cover cloud services that allow developers to code, compile, run, and debug within a browser without installing any software on their computers.

For modern embedded developers, it's not enough to only learn about embedded programming. It's also essential to understand cloud concepts, containers, and automated building and testing. In this chapter and in *Chapter 9, Implementing Continuous Integration*, we will review the ways embedded development projects take advantage of the cloud to automate, simplify, and improve productivity.

For now, in a nutshell, this chapter will cover the following:

- The fundamentals of cloud development
- Coding and containers in the cloud
- A crash course on containers
- Executing software and debugging in the cloud
- Running and debugging
- Getting to know Keil Studio Cloud
- Other cloud development possibilities

The fundamentals of cloud development

Source code creation and management is the fundamental activity of software development. In the past, source code was managed inside a corporate network on a shared server using one of many version control systems. While at the office, developers accessed the source code, and when they traveled or went home, the source code could be accessed using a VPN.

Gradually, the number of version control systems shrank, and developers consolidated on Git, an open source, distributed version control system. Along the way, services such as GitHub started offering a place to store source code and collaborate with other developers.

Today, tens of millions of developers use GitHub to share code and work on software products. In recent developer surveys, more than 90% of all software developers have used GitHub on a project, revealing its ubiquity in software development. In this chapter, we will use GitHub as a starting point for exploring additional ideas in software development. Other platforms, such as GitLab and Bitbucket, are available, which do similar things. Moving your version control system to the cloud is the first step toward cloud development. It's an easy step that saves developers from having to maintain servers, worry about data backups and power outages, and make sure there is enough storage. GitHub facilitates public and private projects and includes options for sharing projects with other GitHub users when needed.

Storing source code in a service such as GitHub enables additional features that developers can use to take advantage of cloud resources. If you are not familiar with Git or GitHub, there are numerous tutorials, videos, and books to learn how they work. Take some time to learn – it's worth it.

An editor, such as a text editor or an **integrated development environment** (IDE), is the second fundamental tool for developers. As source code managers have come and gone, so have editors and IDEs. In the past few years, the developer community has consolidated on **Visual Studio Code (VS Code)**. It has become a popular tool for software development – and for good reason. It supports many languages, runs fast, works on many computers (including Arm computers), and has a strong ecosystem of extensions. VS Code strikes the right mix of being better than a plain editor without having the complexity and specificity of an IDE.

Let's take the dot product example introduced in *Chapter 5*, *Optimizing Performance*, and experiment with some cloud-based development possibilities.

Coding and containers in the cloud

To replicate the example in this section, you will be using the following:

Platform	Raspberry Pi Pico
Software	Dot product
Environment	Gitpod
Host OS	Linux (Ubuntu)
Compiler	GCC
IDE	VS Code

The source code for the dot product example is stored in GitHub and is the same code that we used in *Chapter 5*, *Optimizing Performance*. Here's the link for your reference: `https://github.com/PacktPublishing/The-Insiders-Guide-to-Arm-Cortex-M-Development/tree/main/chapter-5/dotprod-pico`.

In *Chapter 4*, *Booting to Main*, we demonstrated how to clone the GitHub repository on a host machine, install the Pico C/C++ SDK, compile the code, and run it on the Pico. We showed how to do this on a Raspberry Pi 4, but this is possible on other host machines too. This local style of development makes all aspects of software development possible, including coding, compiling, running, and debugging.

Let's see what else is possible with cloud development.

GitHub makes it easy to look at the code from a browser. Developers often spend time on GitHub just looking at the code, reviewing it, or quickly checking something. This is very convenient, as it can be done from any browser. There is an **edit** button to make changes to files in GitHub, but it's a text box, which is not good for programming. It works well for small changes to things such as README files, but not for coding.

GitHub (and other companies) have been introducing features to make coding in the cloud more and more like local software development. This is generally done in a browser. Let's look at some coding in the browser alternatives.

Using VS Code via github.dev

The first capability to code in the browser is to open VS Code. This can be done from a GitHub repository in two different ways:

- Change the GitHub URL from `github.com/some-repo` to `github.dev/some-repo`
- Press the period or full-stop (.) key to jump straight into VS Code

Visit the dot product repository link provided earlier and use one of the preceding methods to open VS Code. It is very smooth. There is no need to clone the repository and nothing is downloaded to your computer. It simply works on any machine with a browser, including a tablet or phone.

Settings, such as the color theme, are saved in your GitHub account and can be used regardless of the computer you use. Some extensions can be installed as well. After a few minutes, you will realize that you can edit code and commit changes, but there is no other storage. Extensions that do things such as formatting, key binding, and implementing themes and views work fine. For example, the vim extension works well. Extensions that need a computer to run something such as compiling code cannot be used. This is confirmed by the fact that there is no terminal available on this `github.dev` VS Code site!

The next level of cloud development is based on the concept of a **development container** (**dev container**). Let's look at how to build and do development using containers.

Dev containers

Containers have become very popular for deploying web applications, such as websites or web-based services. Applications consist of (and depend on) many types of files. Applications are a mix of binary executables, scripts, libraries, and related programs provided by an operating system. Without containers, it can be difficult to confirm whether a particular computer has the required software an application depends on. Containers make it easy to package software in a reproducible way that includes everything an application needs—and only what the application needs.

This means the entire filesystem that an application requires can be bundled together. It also means all the files the application doesn't need don't need to be included because deployment containers are constructed to do one thing – run the application. This keeps containers lightweight and robust at the same time.

Docker is a container build and runtime tool for software developers. There are others, but Docker is the most popular and is the easiest place to start when learning. Docker containers can be thought about in a few ways. One way is that Docker enables building and running an application-specific filesystem. Another way is that Docker is a solution for the "*it works on my machine*" problem, as containers avoid issues such as missing files and different versions of software that cause applications to fail.

If you are not familiar with containers and Docker, look at any of the numerous tutorials available online. There are free, complete courses on YouTube about Docker to quickly get you up to speed. Here is one example that is long but simple to follow and engaging: `https://www.youtube.com/watch?v=3c-iBn73dDE`. Whatever your learning style is, chances are a free tutorial exists in that format.

Dev containers use the container concept. Instead of web application deployment, however, these containers are used primarily to create consistent developer environments. When starting a new project, the first step is to get a new computer, install some software, get the project source code, and try to build and run it. This can take quite some time and it may be difficult to get a working machine for a very complex project. Dev containers aim to avoid this by doing the development in a container that includes all the required software. The new developer is immediately able to start working on the project, only requiring the dev container (and ideally some documentation!).

Now that we understand the difference between dev containers and deployment containers, let's see how they can be used for the Raspberry Pi Pico example.

Dev container services

Multiple services allow dev containers to be used immediately in a browser. These services are amazingly simple but powerful. They use a combination of VS Code and a container in the browser for software development. This addresses the lack of compute we observed with `github.dev` previously.

There are two ways to use dev containers:

- Start from a ready-to-use container that is provided by the service
- Bring your own container with your favorite (or required) software included

Many projects don't require custom containers. This is true if the majority of the software is available in a ready-to-use container and the missing items can easily be added using the Linux package manager or by fetching and installing software from the internet. On the other hand, if the required software is more complex to install, difficult to automate, or available behind a login, then making a custom container is the way to go. We will cover both options.

Two popular dev container services are **GitHub Codespaces** and **Gitpod**. GitHub Codespaces is part of GitHub whereas Gitpod is another service that provides dev containers in the cloud for GitHub projects. This example will highlight how to use Gitpod.

Initializing Gitpod

There are two easy ways to open a GitHub repository in Gitpod:

- Install the Gitpod extension in your browser and use the button in GitHub that opens the project
- Add the `gitpod.io#` prefix to the URL for the GitHub project and Gitpod will open the project

In either case, there is no need to create a new Gitpod account – simply use your GitHub credentials to log in and get started. Open the dot product project for the Raspberry Pi Pico using either method described previously. The following screenshot shows the GitHub repository with the Gitpod button displayed after downloading the Chrome browser extension found here:

```
https://chrome.google.com/webstore/detail/gitpod-always-ready-to-co/
dodmmooeoklaejobgleioelladacbeki
```

To try out GitPod, use the GitHub site for this book:

`https://github.com/PacktPublishing/The-Insiders-Guide-to-Arm-Cortex-M-Development`

If you visit GitHub and have the browser extension installed, click the Gitpod button.

The direct link to open Gitpod is here:

`https://gitpod.io/#https://github.com/PacktPublishing/The-Insiders-Guide-to-Arm-Cortex-M-Development`

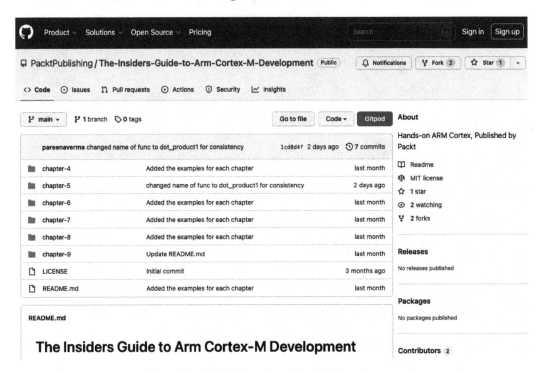

Figure 8.1 – A GitHub project with the Gitpod button

Opening the GitHub site using the preceding link sets up all of the needed tools for the Raspberry Pi Pico dot product example. This is done using the `.gitpod.yml` file at the top level of the GitHub repository.

To instruct Gitpod to install and build the SDK when creating a workspace, look at the `.gitpod.yml` file. It is a hidden file and may not be viewable in the Gitpod space but can be seen in GitHub at the top of the files listed. Gitpod reads this file when a new workspace opens. Here are the `.gitpod.yml` file contents:

```
tasks:
  - name: build sdk
    init: |
      pushd /workspace
      sudo apt purge -y --auto-remove cmake
      wget -O - https://apt.kitware.com/keys/kitware-archive-
latest.asc 2>/dev/null | gpg --dearmor - | sudo tee /etc/apt/
trusted.gpg.d/kitware.gpg >/dev/null
      sudo apt-add-repository 'deb https://apt.kitware.com/
ubuntu/ focal main'
      sudo apt update
      sudo apt install -y cmake
      pushd ~/
      wget https://github.com/ARM-software/CMSIS_5/archive/
refs/tags/5.9.0.zip && unzip 5.9.0.zip && rm 5.9.0.zip && ln
-s  CMSIS_5-5.9.0 CMSIS_5
      popd
      wget https://raw.githubusercontent.com/raspberrypi/pico-
setup/master/pico_setup.sh
      chmod +x /workspace/pico_setup.sh
      SKIP_VSCODE=1 SKIP_UART=1 ./pico_setup.sh
      popd
      source ~/.bashrc
      cd chapter-5/dotprod-pico
```

There are different types of tasks that can be defined; building the SDK is a one-time event for each new workspace, so it uses the `init` tag and only runs on workspace initialization. There is also a `command` tag that runs each time a workspace is opened. This example only contains initialization tags to set up the SDK, and when the project opens, the commands in `.gitpod.yml` will be processed. There are some operating system updates and then the SDK is installed and compiled.

Every time a new workspace is created you will see the system updates and SDK installation happening in the terminal. After a few minutes, the SDK is set up and ready to use. The following screenshot shows a fresh workspace being created with these defined tasks automatically running:

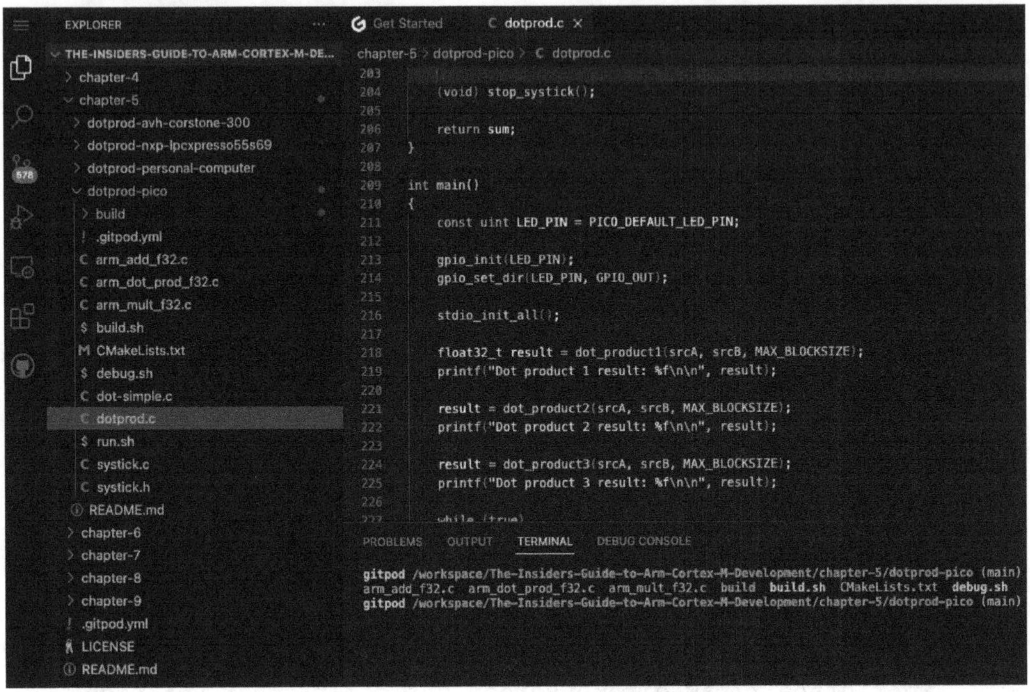

Figure 8.2 – The Gitpod workspace

At this point, the dot product example project from *Chapter 5*, *Optimizing Performance*, is ready to compile. Note that by default, Gitpod clones repositories into the /workspace directory, where the GitHub repository is now located.

These are the steps to build the project after the automatic workspace initialization is complete:

```
mkdir build ; cd build
cmake ..
make
```

Now, the dot product software is built and ready to be loaded onto the Pico board. Later in the chapter, we will cover how to connect our Pico board (on our desk) to this Gitpod workspace (in the cloud) to actually run and debug our software interactively. First, let's look into more details on the Gitpod workspace and how to get more value from it via customization.

Workspace images

Gitpod uses a default container for the workspace called `workspace-full`. The description of this container is available on GitHub, alongside other container options. Here is the `workspace-full` description: `https://github.com/gitpod-io/workspace-images`.

One of the drawbacks of the demonstrated approach using Gitpod is that the SDK needs to be installed each time a new workspace is created. This workflow is good if the SDK is frequently changed and you want to use the latest version. The workflow can be improved if the SDK (or other dependencies in your project) is fixed for the length of the project. Also notice that `pico_setup.sh` is installing additional Linux packages. This means the Linux packages will be installed every time a new workspace is created.

The solution to re-installing software for each new workspace is to create a custom container image with the required software. It also solves the problem of installing software that is not easy to automatically download and install. An example is software that requires a browser login to download.

Creating a custom Docker image with the SDK and other Linux packages already installed saves developer time (assuming Gitpod can start a custom container faster than the prebuilds take to run). Now, when the workspace is started, the dev container will have everything initialized and ready to go.

Before we deal with creating a custom container for our example Gitpod project, let's take a step back and look at containers in general. We will show how to create a dev container for the Pico project that can be run on any computer with Docker installed, including Windows, macOS, and Linux. After that, we will make a custom dev container for Gitpod and see how to use it in Gitpod to save setup time.

A crash course on containers

Creating a dev container for the Pico project gives developers the freedom to work on any computer with Docker installed. This can be Windows, macOS, and Linux. The same container will run the same regardless of the operating system it is used on.

Installing Docker is a straightforward process on any operating system, so refer to the main Docker documentation for the most up-to-date instructions for whatever operating system you are using locally to test this concept. Here is a link to the instructions: `https://docs.docker.com/engine/install/`.

Windows and macOS users can easily install Docker Desktop by downloading and running the installer.

Installing Docker on Ubuntu Linux is straightforward and can be done from the command line. Use the following instructions to download, install, and run an example Docker container, `hello-world`, to validate that it is working as expected:

```
sudo apt update
sudo apt upgrade -y
```

```
curl -fsSL get.docker.com -o get-docker.sh && sh get-docker.sh
sudo usermod -aG docker $USER ; newgrp docker
docker run hello-world
```

Creating a Dockerfile

We are going to create two dev containers for the Pico SDK to use for development. First, we will create a base container with the Pico SDK and other software needed to build the dot product example. From the first container, we will then add a VS Code server, so we can use it from a browser to edit files. This is optional for developers who want to have VS Code built into the dev container and accessible from a browser. Developers who prefer to attach their own VS Code or use the command line can use the first container.

Dev container creation starts from a Dockerfile. This file contains all the commands needed to create the content of the container. The first step is to select a base container, which generally means selecting the base filesystem. This can be an operating system such as Debian or Ubuntu or a runtime such as Python. This first line of the Dockerfile sets the initial filesystem, which will be in the container. Everything after that is an addition to the filesystem needed to run the desired development tools and application.

For this project, we are going to use Ubuntu 22.04 as the base image. The Pico SDK requires a Debian-based distribution because the `pico_setup.sh` script uses `apt` commands to install extra software. The Pico SDK also requires a pretty new version of `cmake`, so I selected Ubuntu 22.04 to get a new-enough version of `cmake` for the SDK to work.

We will review the commands listed in the Pico dev container Dockerfile, which you can review here in detail: `https://github.com/PacktPublishing/The-Insiders-Guide-to-Arm-Cortex-M-Development/tree/main/chapter-8/pico-dev-container`.

The first line of the Docker file is as follows:

```
FROM ubuntu:22.04 as base
```

This `FROM` line also uses `as base`, which allows us to reference this container when we build the second container. Building two containers from one Dockerfile is called a multi-stage build since we are going to build two containers and the second one will add on or take content from the first. This is common practice when building applications. The first container builds the software and the second container copies the build results to the second container. This makes it possible to just include the needed runtime files and leave all the intermediate build files behind.

Farther down in the Dockerfile, we use the following:

```
FROM base as vscode
```

This creates the second image, which has a target name of vscode and inherits from the first image, named base.

The remainder of the Dockerfile consists of commands to add more files to the container.

There are a few statement types in the Dockerfile. The statements are also known as instructions. The details of each instruction type are in the Docker documentation and covered by tutorials. Here is a short summary of each:

- RUN will execute the commands that follow and save the resulting image.
- ENV is an environment variable that can be used during build and run. ENV values can also be overridden (or changed) during future runs.
- ARG is like an environment variable, but is only used during the build stage and doesn't persist in the final image. ARG values are good for specifying things during the build.
- WORKDIR sets the working directory for future RUN (and other) commands.
- USER sets the Linux username of the user that will run the future RUN commands.
- EXPOSE informs Docker to listen on a specified port at runtime.
- ENTRYPOINT is an instruction to start a program automatically when the container is run.
- The # symbol is a command in a Dockerfile.

With that background information in hand, and an overview of the Dockerfile creating our base and vscode images, let's build it.

Building and storing the containers

The docker build command is used to turn the Dockerfile into a container image.

The build command needs a few things to run:

- The name of the Dockerfile
- The name of the image to create
- Any additional arguments, such as values for the ARG variables in the Dockerfile
- The stage to build if the Dockerfile has multiple stages

Take a look at the build.sh script in the pico-dev-container project.

The first container is built using a command:

```
docker build --build-arg "CMSIS_5=5.9.0" -t pico-dev:latest
--target base -f Dockerfile .
```

It specifies the name of the container and which target in the Dockerfile to build. The last part specifies to use the Dockerfile from the current directory.

The second container is very similar. There are several additional build arguments for the version, files, and paths of VS Code, so please use the `build.sh` script in the GitHub repository to build it.

The container name is made of two parts. The first part is `pico-dev` in both cases. The second part after the colon is the tag. I think of it more like the version. In this case, both images can have the same name but are differentiated by the tag. When they are stored in a container repository, they can be managed together in one repository.

Docker images make it easy to transfer the dev containers from one machine to another. Instead of having to copy files across machines, container repositories provide storage. Containers can be built on a computer and transferred to a repository to be stored. This is known as a *push*. Then, another computer can download the same container image and immediately run it. This is known as a *pull*. Many repositories store images, and you can even create your own, but the most common is Docker Hub. To learn now to push and pull images, create an account on `https://hub.docker.com` and look over their documentation. Using Docker Hub for storage is much easier than copying files directly from computer to computer.

Building arguments and multi-architecture support

If you study the `build.sh` file closely, you will notice two other details. First, the build arguments are passed to `docker build` using `--build-arg`. This allows the ARG values in the Dockerfile to be set when `docker build` is run. The ARG values specify the versions of software to use in the build and filename, which may change as newer versions become available. In the Dockerfile, we have the version of the CMSIS software to use and the file to use for the VS Code server. This way, when versions change, the `build.sh` script can be updated and the same Dockerfile continues to work.

The second thing to notice is the multi-architecture support. Historically, desktop and laptop computers, as well as servers, used Intel or AMD processors. The architecture may be called x86, x64, or amd64. More recently, the Arm architecture has gained market share in laptops and servers. When the architecture (or instruction set) is different, the software compiled must be compiled for the architecture of the computer. We want this dev container to work on both architectures so developers can select the computer they want to use and it will just work. The `pico-dev` containers work on any computer: a Raspberry Pi 4, a macOS laptop, a Windows computer with either architecture, or a Linux server with either architecture.

There are numerous ways to create multi-architecture containers. The most common is the `docker buildx` command. If you are interested in `buildx`, there are numerous tutorials on how to use it with either instruction set emulation or multiple computers to build for each architecture.

To keep things easy for our example, I (Jason) commonly use another technique to create the multi-architecture container: the `docker manifest` command. You will see the `build.sh` file that the architecture of the computer is detected using the `uname` command and that is added to the tag. For example, see the following:

```
-t pico-dev:latest-$(arch)
```

There is a script named `push.sh` that uploads the two container images for that computer to Docker Hub with the architecture in the tag name. When both the `x86_64` and the `aarch64` images are uploaded, the `docker manifest` command joins them into a single image without the architecture in the tag name. This image is a multi-architecture image. When the image is retrieved using `docker pull` or `docker run`, the downloaded image is correct for the architecture of the computer.

Look at the `join.sh` script to see how it combines the images for each architecture into a single multi-architecture image. There is a `create` command to join them:

```
docker manifest create jasonrandrews/pico-dev:latest \
--amend jasonrandrews/pico-dev:latest-aarch64 \
--amend jasonrandrews/pico-dev:latest-x86_64
```

There is also a `push` command to send to Docker Hub:

```
docker manifest push --purge jasonrandrews/pico-dev:latest
```

To see this working in action you can use `docker pull` on any computer to download the `pico-dev:latest` image with this command:

```
docker pull jasonrandrews/pico-dev:latest
```

Docker will automatically know which image is the right one for the computer and download that one. This is the benefit of the multi-architecture image. The same commands can be used on a computer of either architecture and Docker knows to retrieve the correct one.

Running the dev containers

The `docker run` command is used to run a container on any computer which has Docker installed. Images already on the computer can be run or images can be automatically downloaded from Docker Hub as needed.

To run the base container on any computer you are using (with Docker), simply run the following command:

```
docker run -it jasonrandrews/pico-dev:latest
```

Putting my (Jason's) Docker Hub username on the front of the image name tells Docker to download it from Hub and run it. If you have a local image on your computer, it may not have the username in front of it. You can refer to the `push.sh` script to see how the `docker tag` command was used to add the username of the Hub account before uploading it and can customize it to your own account if desired.

When `pico-dev:latest` is run, you will come to a Linux shell prompt:

```
ubuntu@b75cfd449087:~$ ls
CMSIS_5  CMSIS_5-5.9.0  pico  pico_setup.sh
```

Clone the dot product project from GitHub:

```
git clone https://github.com/PacktPublishing/The-Insiders-
Guide-to-Arm-Cortex-M-Development
```

Now, follow the same process we used in *Chapter 6*, *Optimizing Performance*, to build the dot product example:

```
cd The-Insiders-Guide-to-Arm-Cortex-M-Development/chapter-5/
dotprod-pico
mkdir build ; cd build
cmake ..
make
```

Over the past few sections, we have demonstrated how to use a dev container to build our provided dot product example. Within a few minutes, anybody can download the container to any computer with Docker installed, clone the dot product repository, and build it. There is no need to install operating system packages, run the `pico_setup.sh` script, or install CMSIS_5 source code. This ease of use is central to the value that working with the cloud offers.

Integrating dev containers with VS Code

There are multiple ways to edit the code in the dev container. Certainly, vim can be used in the terminal to edit code, but let's look at a few ways to use VS Code with the `pico-dev:latest` container:

- VS Code outside the dev container

  I. Run the dev container from the previous section (if it's not already running):

  ```
  docker run -it jasonrandrews/pico-dev:latest
  ```

Separately, open VS Code locally on your computer. Install the Docker extension. Using that Docker extension, find the running container. Now, use the VS Code remote explorer to connect to the running container. You can find detailed instructions on this process in the VS Code documentation on containers: `https://code.visualstudio.com/docs/containers/overview`.

Once connected, open the `dotprod-pico/` folder. Now, VS Code can be used to edit, compile, and access the terminal of the container. While it looks and feels as though you are developing inside a local machine, you are actually working inside of the dev container.

- VS Code inside the dev container:

Recall that we previously built two dev containers. The second one has the `vscode` tag and includes a VS Code server inside of it. This is what we will demonstrate now. Run it with this command:

```
docker run -it -p 3000:3000 jasonrandrews/pico-dev:vscode
```

Notice the extra `-p` argument to open port `3000` for the browser connection to the VS Code server. When the dev container starts, you will see a message:

```
Web UI available at http://localhost:3000/
[02:47:37] Extension host agent started.
```

Using a browser, connect to the VS Code server at `http://localhost:3000`, and you should see VS Code in the browser. The `dotprod-pico` repository is already there because we installed it into the dev container. Look back at the Dockerfile if needed. Open the `dotprod-pico` folder – you can now edit the source code again and use the terminal to build the application. Note that this port forwarding method does not create a secure connection, but it can be leveraged to communicate across multiple machines. To create a secure connection, a reverse proxy such as `nginx` can be used, but this is beyond the scope of this book. If interested, you can Google `nginx reverse proxy docker` and look into some recently published instructions.

- VS Code inside the dev container from another machine:

If the dev container is running on another machine, such as a Raspberry Pi 4 on my network with the IP address 192.168.68.96, I can still use the browser to connect. Use `ssh` to connect to the Pi with port forwarding, use `http://localhost:3000` again, and see the VS Code open. Now, I'm working on the dev container on my Raspberry Pi 4, but the container is exactly the same.

Here is the `ssh` command with port forwarding:

```
ssh -L 3000:localhost:3000 192.168.68.96
```

We now have some good experience building and running dev containers on different types of computers. To reiterate, the `pico-dev` container can be run on any machine with Docker, from the Raspberry Pi 4 to a cloud server. The machines can have Arm or x86 architecture.

Now, let's circle back to Gitpod and see how to make a custom Docker image for use in Gitpod.

Executing software and debugging in the cloud

Making a custom docker image for Gitpod is not much different from the generic dev container. The main difference is the starting point for the image. Instead of starting from Ubuntu, we start from the Gitpod base container, `gitpod/workspace-base`. The second difference is the username. We created the ubuntu user previously, but with the Gitpod container, the username is already set to `gitpod`. The final difference is that the package installation requires `sudo` in front of the commands because everything in the Dockerfile is run by the `gitpod` user.

Creating a custom Gitpod image

Take a look at the Dockerfile in the `pico-dev-container` GitHub repository (`https://github.com/PacktPublishing/The-Insiders-Guide-to-Arm-Cortex-M-Development/tree/main/chapter-8/pico-dev-container`) to see the differences in the Dockerfile intended for Gitpod. Gitpod is only available for the x86 architecture, so there is no `aarch64` image for Gitpod. Maybe someday, we hope, Gitpod will add support for the Arm architecture.

The Gitpod container is on Docker Hub with the other dev containers we have used in this chapter. It uses the `jasonrandrews/pico-dev:gitpod` Gitpod tag.

To configure Gitpod to use the custom container, add it to `.gitpod.yml` for any GitHub project. We have done this in the provided `.gitpod.yml` at the top of the GitHub project for the book.

The first line changes from using the tasks discussed in the previous section to using a custom container image. This is the only line needed for the custom container image.

```
image: jasonrandrews/pico-dev:gitpod
```

Edit the `.gitpod.yml` file and remove the # symbol from the first line and comment (using #) or delete the remainder of the file. The easiest way is to fork the book repository into your own GitHub account, edit the `.gitpod.yml` file, and launch Gitpod.

With the custom container in use, the previous commands present in the `.gitpod.yml` file that install the Pico SDK, add `CMSIS_5`, and change the version of `cmake` are not needed! You can comment out all these tasks by adding a # symbol in front of the lines if desired; the only line necessary in `.gitpod.yml` now is for referencing the `pico-dev:gitpod` container image.

If you make this `.gitpod.yml` file change and start a new Gitpod workspace, the new Gitpod workspace is ready to work right at startup and requires no additional tool installation. Either strategy (the custom container or the `init` commands in `.gitpod.yml`) can work. You will have to decide which approach works best for your specific project.

With a custom container in place, we have all the tools to build, run, and debug the dot product application.

Running and debugging

Gitpod works well for coding and building microcontroller software, but what about running and debugging it? This is more complex by nature since the Raspberry Pi Pico is on our desk and not physically attached to the Gitpod cloud instance.

Let's take the scenario where the Raspberry Pi Pico is on my desk and connected to the Raspberry Pi 4. In *Chapter 4, Booting to Main*, we reviewed how to use `openocd` and `gdb` to connect, load, and run the `hello world` program on the Pico. We will use some of these concepts here as we work in Gitpod and use `gdb` to connect to and debug the dot product application. Review *Chapter 4*, if needed when following the next few steps:

1. First, return to your Gitpod workspace – created either from the custom container image or the initialization commands – and build the application with debug settings:

    ```
    mkdir build ; cd build
    cmake -DCMAKE_BUILD_TYPE=Debug ..
    make
    ```

2. From a terminal on your local Raspberry Pi 4, use `ssh` connect to the Gitpod workspace. There are multiple ways to do this: you can add your own public `ssh` key or use an access token. The second method is easiest to implement on the first try. To do so, go to the `gitpod.io` dashboard by clicking on the three lines on the top-left-hand side of the Gitpod VS Code workspace and selecting **Gitpod: Open Dashboard**. In the `gitpod.io` dashboard, click on the three dots on the right-hand side of the workspace and select **Connect via SSH**:

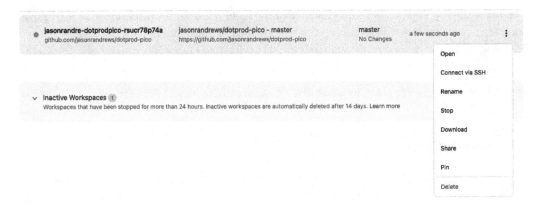

Figure 8.3 – The Gitpod workspace list

3. Click on the **Access Token** tab and copy the given command. Paste this command into a terminal on your Raspberry Pi 4. The only necessary addition is to set up reverse port mapping for port 3333. This provides the gdb to the openocd connection.

    The command will look as follows (with your access token values between the quotations):

    ```
    ssh -R 3333:localhost:3333 'jasonrandre-dotprodpico-rsucr
    78p74a#WSC41GSOcCnRh_2IfRrk8nYWtTEIDwCY@jasonrandre-
    dotprodpico-rsucr78p74a.ssh.ws-us54.gitpod.io'
    ```

    Notice -R 3333:localhost:3333 is the required added part of the command enabling reverse port mapping.

4. Now, run the command in the debug.sh file (available in the same example on GitHub) from the chapter-5/dotprod-pico/ directory on the Raspberry Pi 4. openocd will sit waiting for a gdb connection now.

5. Make the gdb connection from Gitpod using the same gdb command sequence from *Chapter 4*. This time, however, Gitpod will automatically connect back to your local Raspberry Pi 4 and debug the code on the Pico. The final command of just the letter c will continue running the application until any specified breakpoints – in this case, at the main function:

    ```
    gdb-multiarch build/dotprod.elf


    (gdb) target remote localhost:3333
    (gdb) load
    (gdb) b main
    (gdb) c
    ```

In summary, Gitpod offers a very powerful environment to code, compile, and debug in the cloud. It can also be somewhat complex for non-Linux users, as you may have noticed along the way. Many of the operations require command-line knowledge that does not overlap too much with the traditional required embedded development skills, operating from a straightforward GUI-based IDE that simplifies this complexity for you.

In the next section, we look at Keil Studio Cloud, a cloud-based IDE that offers GitHub integration, debugging features, and support for many Cortex-M development boards. It provides a simplified alternative to the highly customizable Gitpod flow we have explored already. Keil Studio Cloud also provides a built-in compiler service and improved debugging experience, which is attractive to many developers.

Getting to know Keil Studio Cloud

To replicate the example in this section, you will be using the following:

Platform	NXP LPC55S69-EVK
Software	Blinky
Environment	Keil Studio Cloud
Host OS	Any
Compiler	Arm Compiler for Embedded
IDE	Keil Studio Cloud

For this section, we take a break from the Raspberry Pi Pico and go back to the NXP LPC55S69-EVK which uses the Cortex-M33 processor.

To get started with Keil Studio Cloud, visit `https://keil.arm.com`. Select the **Find your Hardware** button and search for `55S69`. The LPCXpresso55S69 will show up, as shown in the following screenshot:

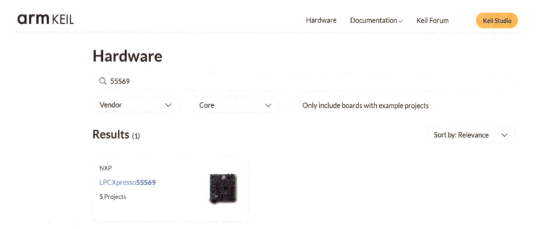

Figure 8.4 – Hardware selection on Keil Studio Cloud

After selecting this board, you can choose the *Blinky* simple example and open it in Keil Studio by selecting the **Open in Keil Studio** button. You will be redirected to `studio.keil.arm.com` with this example project ready to edit. You may need to create an account to continue, which is free.

It provides a browser-based editor, built-in compiler service, and debugger for building and deploying your embedded software projects. It also provides source control integration with Git, so you can easily perform common Git actions and push your project directly from the Keil Studio IDE to GitHub.

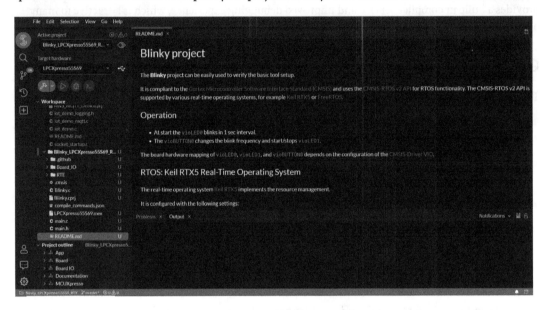

Figure 8.5 – The Keil Studio Cloud user interface

On opening the Blinky project in Keil Studio, your workspace on the left-hand side gets populated with the project source files and you can view or make changes to them using the built-in editor.

Next, onto building your software for the NXP LPCXpresso55S69 target. This leverages the hosted compiler service and builds your binary image using the Arm Compiler for Embedded toolchain. To load the binary and either run or debug it on the board, first, connect your board to the machine running Keil Studio Cloud, and then from the browser, select the USB icon next to the target hardware. If you have problems pairing to CMSIS-DAPLink, likely, the CMSIS-DAP firmware for the debug unit that connects the Debug Port to USB needs to be updated. Follow the steps outlined here to update the DAP firmware: `https://os.mbed.com/teams/NXP/wiki/Updating-LPCXpresso-firmware#binary-downloads`.

When your board is successfully connected, the USB icon will turn green, and if you hit the **Run** button, your compiled binary is downloaded onto your board, and you should see the blinking LED. The debug option is simple and gives you the basic features to step through your code, set breakpoints, and understand program flow execution.

It is exciting to see new tools such as this where you can just launch a browser, develop your microcontroller code, and flash it to the board on your desk.

Other cloud development possibilities

There are other tools and environments possible to leverage in cloud-based development. This section covers some other options that you may want to consider for your projects, or at least be aware of.

Cloud virtual machines

Another way to develop in the cloud is by using virtual machines. We saw in *Chapter 4* and *Chapter 5* how to use an **Amazon Machine Image** (**AMI**) from AWS Marketplace for development. We learned how to use ssh to connect to the AWS EC2 instance and run applications such as the machine learning examples on the Corstone-300 Fixed Virtual Platform.

One thing we didn't cover is using VS Code in the AMI for development work. There are two common ways to use VS Code with a virtual machine.

One way is to use an ssh connection. The VS Code Remote SSH extension can make the connection to a remote virtual machine. Various articles for connecting VS Code to a remote machine via ssh are available, and we recommend searching for ssh to remote machine VS Code for the most recent information. There are pointers to resources in the *Further reading* section of this chapter as well.

A second way to use VS Code in the Arm Virtual Hardware AMI is to use the built-in VS Code server. Add port forwarding for port 8080 to the ssh command and the VS Code server is available via the browser.

Use this command to use ssh to connect from your local machine to the EC2 instance with port 8080 forwarded:

```
ssh -i <your-private-key>.pem  -L 8080:localhost:8080
ubuntu@<your-ec2-IP-address>
```

Once in the EC2 instance, use a browser on your local computer and connect to http://localhost:8080:

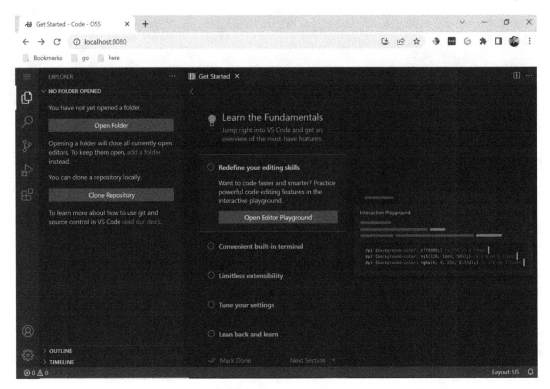

Figure 8.6 – VS Code with a ssh connection

You will see a fresh VS Code in the browser, ready to be configured. You can use it to develop code directly on the EC2 image.

Virtual desktop in VM

Sometimes, graphical tools are required for development and a Linux desktop is helpful or required. The Arm Virtual Hardware AMI also provides a virtual desktop via **Virtual Network Computing** (**VNC**). We will start the EC2 instance again and connect to it using ssh, but this time, forward port 5901. Use this command to connect using ssh and forward port 5901:

```
ssh -i <your-private-key>.pem  -L 5901:localhost:5901
ubuntu@<your-ec2-IP-address>
```

After connecting, run these commands (you will have to set a password for VNC after running the first command):

```
vncpasswd
sudo systemctl start vncserver@1.service
```

The second command will start the VNC server on the EC2 instance. To connect to this server, you can install a VNC client on your local computer (we recommend TigerVNC or TightVNC) and connect to `localhost:5901` with the VNC client:

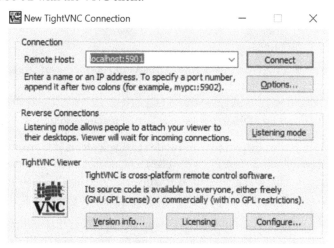

Figure 8.7 – Making a VNC connection

Your virtual desktop from the AMI should then appear. With VNC running, you can install VS Code and any other graphical applications on the virtual machine and run them:

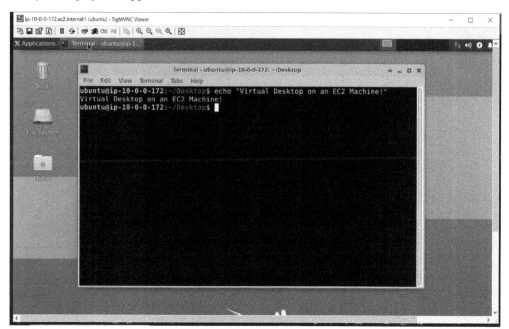

Figure 8.8 – A Linux desktop using VNC

Cloud virtual machines can be used in multiple ways. We have demonstrated `ssh`, VS Code in the browser, and VNC for a Linux desktop. Projects may use one or more of these depending on the tools required.

Summary

This chapter walked you through the various ways that cloud services can streamline Cortex-M software development. We investigated how to effectively use a virtual machine in the cloud as a remote computer with offerings such as Gitpod and EC2 instances. VS Code was highlighted as the emergent cloud-based IDE, compatible across various virtual machine solutions.

Next, we reviewed how to make environment and code sharing easier and more consistent through dev containers. Docker containers are the most popular technology, and an example of how to build, share, and run a Docker container was presented. We further analyzed how to create multi-architecture containers that can run seamlessly across any architecture and operating system that can run Docker.

Finally, we showcased an example cloud-based development process that still ran and debugged code on a local embedded board. The presented tools and methodologies in this chapter unlocked the advantages of cloud-based development, such as environment consistency and ease of use, for embedded projects.

New cloud services continue to come out every year and these rapid iterations are likely to continue. While this chapter provided an overview of how to leverage existing services, more possibilities to increase development efficiency are certain to emerge. The last chapter of this book, *Chapter 10, Looking Ahead*, presents some nascent cloud technologies at the time of publication in 2022. Becoming comfortable with developing embedded software in the cloud is a surefire way to add value to your future projects.

Another methodology that has steadily become the best practice for software development is automatic software testing. This practice is referred to as continuous integration, which can drastically increase software quality and reduce development time when implemented correctly. The next chapter describes the basics of continuous integration and details several ways to implement it effectively in embedded projects.

Further reading

To learn more about the topics that were covered in this chapter, take a look at the following resources:

- Gitpod getting started guide: `https://www.gitpod.io/docs/getting-started`
- An overview of Docker: `https://docs.docker.com/get-started/overview/`
- Docker reference documentation: `https://docs.docker.com/reference/`

- Docker general cheat sheet: `https://dockerlabs.collabnix.com/docker/cheatsheet/`

- Dockerfile instructions cheat sheet: `https://devhints.io/dockerfile`

- An overview of GitHub Codespaces: `https://docs.github.com/en/codespaces/overview`.

- An overview of the Gitpod `.yml` file, to learn how to initialize Gitpod environments: `https://www.gitpod.io/docs/config-gitpod-file`

- An overview of Docker Hub: `https://docs.docker.com/docker-hub/`

- Multi-architecture building with `docker manifest` and `buildx`: `https://www.docker.com/blog/multi-arch-build-and-images-the-simple-way/`

- An overview of VS Code remote development over SSH: `https://code.visualstudio.com/docs/remote/ssh-tutorial`

- An overview of TigerVNC: `https://tigervnc.org/`

9

Implementing Continuous Integration

IoT and ML have brought about an explosion of intelligent connected devices that bring massive benefits to various sectors such as industrial, healthcare, and many others. However, with interconnectivity and intelligence comes a steep increase in the complexity of the software.

Factors such as security, over-the-air updates, and networking stacks are essential for connectivity. To enable ML, models may need to be refreshed multiple times a week to stay accurate. For rich OSs such as Linux or Windows, many of these complications are resolved at the OS level and abstracted from the application running on it. For embedded devices, this is typically not true, and that complexity is passed on to the software developer to sort out.

This step change in complexity for embedded devices is causing a major disruption to how embedded software is managed throughout its life cycle. A *traditional* embedded device, such as a washing machine from the 1990s, has its functionality frozen after manufacturing and installation. It is not connected to the internet and no further updates can be delivered. A *smart* embedded device, such as a washing machine from the 2020s, is connected to the internet and often receives updates to enhance functionality, performance, and security over time. This means modern software developers are not done with a project after its initial creation and must support that device software for months and years afterward.

The old, manual embedded software development flows are good enough for supporting *traditional* devices, but the new *smart* devices require a new, smart development flow. This chapter discusses that new flow, which is called **continuous integration** (**CI**). We will discuss the following:

- Understanding the value of CI
- Why embedded software CI can be challenging
- Examples of CI flows

Understanding the value of CI

CI is a best-practice framework for software development that aims to improve software quality while reducing development time and cost. It can best be described by contrasting it with the typical embedded software development flow, which we call *desktop development*.

In the desktop development situation, a team of software developers all works on the same code base at the same time, typically using GitHub or a private code repository. As each developer works, they send their modified code through their own non-standardized custom suite of tests with the hardware board on their desk, about once per day. This code is merged into a shared branch with their team. Developers then pull the shared branch back to their local environment to obtain changes from their team. At this point, the code may or may not be in a working state, and it is sometimes hard to identify the source of bugs that have been introduced by the interaction of different changes.

This flow may sound familiar to you, even if it doesn't match your experiences exactly. It has been the most common way to develop embedded software for the past few decades, though it comes fraught with challenges. These include the following:

- **Merge conflicts**: Occur when two developers try to combine non-compatible code to the shared repository.

- **Hard-to-fix bugs**: Occur when recently committed code breaks the software in an unknown or complicated way.

- **Duplicated efforts**: Occur when commits are not done often – developers can solve the same problem and waste time.

- **Inconsistent metrics**: Occur when there is no evidence of code quality through robust test analysis.

- **Near-release chaos**: Occurs when everyone tries to commit their slightly incompatible code versions to the main branch, resulting in many problems.

- **Production errors**: Occur when bugs from development make it into the end product, causing a myriad of unfortunate side effects.

Implementing a CI flow can avoid or lessen the impact of these challenges. In the CI situation, a team of software developers still works on the same code base at the same time. However, each developer commits and merges small code changes to their code branch about every hour, not every day. On this commit, a CI process automatically starts building the full code base to run a standardized regression test suite. As the code passes each stage of testing, it can be pushed to checkpoint branches to automatically keep a trail of working code. After these tests, the changed code is automatically merged with the development branch. This ensures that this branch is always in a working state and up to date. Developers then pull this code back to their local environment and continue working.

This flow addresses the challenges listed earlier in the following ways:

- **Merge conflicts**: Due to the frequency of commits and pulls from the main branch, these conflicts are minor and simple to fix.

- **Hard-to-fix bugs**: Code changes are uploaded in small increments, making changes and revisions easier.

- **Duplicated efforts**: Frequent commits identify areas of overlapping efforts quickly.

- **Inconsistent metrics**: Metrics are automatically generated and recorded for every code commit.

- **Near-release chaos**: Everyone's code base is compatible during the whole development process.

- **Production errors**: CI enforces frequent testing and an understanding of code coverage, helping to prevent these bugs.

In addition to avoiding these issues, having a robust CI flow leads to higher code quality and faster development time. It seems as though everyone should be doing it! Unfortunately, implementing CI flows for embedded software has historically been more difficult than CI flows for a website or a Windows, Linux, or macOS application. The next section discusses these challenges and what can be done to mitigate them.

Why embedded software CI can be challenging

Creating software gets harder when you *develop* on a platform different from the platform you *deploy* on. When developing a website that will be deployed on a Linux-based server, developing on a Linux machine makes it easier to validate software behavior, as development tests will match the production environment. You could use a Linux laptop or a virtual machine to replicate this environment locally. Further, it is easy to access rich OS machines en masse via cloud platforms that offer Linux, Windows, and macOS operating systems.

By contrast, we still develop embedded projects on laptops and PCs but deploy them onto totally different hardware. Developing a smart lightbulb powered by an Arm Cortex-M7 on a Windows laptop is challenging due to this mismatch of platforms. The mismatch requires cross-compilers and, the primary issue, dedicated testing hardware.

It is relatively trivial to spin up dozens or hundreds of Linux instances in the cloud that can run tests for a website under development. There is no problem accessing these platforms and cloud providers have made it easy to dynamically pay for what you need to scale effectively. This advantage does not extend to embedded software developers.

For individuals or very small teams developing software for an embedded device, it may be enough to have a few hardware boards lying around that can test software functionality every day or so. For most embedded software projects, however, the need to validate software through suites of regression tests in a CI flow on hardware is a major challenge.

The most common solution is a local **board farm**. This farm contains dozens or hundreds of development boards that can run a lot of tests in a reasonable time. Having this does enable CI development at scale, but comes with two large drawbacks:

- **Maintenance**: Keeping even dozens of boards in a working state and ready to test with is no small task. Boards can wear down after too many flashes, can malfunction, needing a physical reset to be pressed, or require large rooms to store them, which may require cooling. Moreover, even organizing that many boards becomes a challenge at scale.

- **Cost**: This setup and maintenance can get very expensive. Supporting software for years requires running tests and keeping older hardware around in addition to newer projects, making costs increase over time.

It is often a hard sell to implement a CI flow for embedded software with these significant problems, despite the added benefits. There is an alternative solution for embedded developers to see the advantages of server-based tests and that is with virtual hardware.

Replacing board farms with virtual farms

In *Chapter 3, Selecting the Right Tools*, we introduced an alternative platform for running embedded software: virtual platforms. We have been leveraging **Arm Virtual Hardware** (**AVH**) throughout this book and it is an excellent platform for implementing CI flows at scale. Because AVH runs in the cloud, it is simple to scale up and down.

An important caveat to this solution is that virtual platforms are limited by what they model. There are AVH systems that represent all Cortex-M CPUs, but they will not represent your exact hardware. Specific peripherals you are leveraging, such as Bluetooth, Ethernet, Wi-Fi, sensors, and displays will likely not be the same between the AVH system and your end device. This does not invalidate the advantages of using AVH for CI tests, but it does require an understanding of what software to test where.

> **Important note**
>
> Some AVH systems are starting to represent the software behavior of physical development boards almost exactly, including peripherals, displays, and communication. The Raspberry Pi 4 is modeled in this way. The scope of development boards available is limited at the moment but may become more extensive over time.

All tests are not made equal. We can split them into three different categories that differ in what the software test is trying to accomplish:

- **Unit tests**: Intended to test each individual software component in the smallest chunks possible. These often test individual functions in isolation to ensure they give the correct output with the right inputs. They immediately detect broken code in isolated areas and identifying errors at this level is much easier than in broader tests. They can number hundreds or thousands of tests (sometimes tens of thousands in massive software projects, such as for a self-driving car).

- **Integration tests**: Intended to validate that the interactions between components are working as expected. Errors can occur in communication between systems, functions, or code units, and these tests identify these problems. They can number dozens or hundreds of tests.

- **System tests**: Intended to verify that the complete software system is working as expected. They are also referred to as *black-box* testing, where the inner workings of the system are not evaluated, only the result. These are often used to ensure the system meets project requirements. Often a dozen or fewer of these tests are created.

Regression suites primarily consist of atomized unit tests, with integration tests and systems being far fewer in number. These unit tests validate the functionality of small chunks of code that often center around processing capabilities and don't require the entire system to run effectively. This combination makes them the perfect candidate for running on virtual platforms, which are functionally accurate for the areas they represent. Integration tests, depending on what they cover, can be run either on virtual platforms or physical hardware. System tests should always be run on physical development platforms, or the end device if possible, to be as accurate as possible to the end system.

The most successful embedded CI flows will have a large number of virtual platforms running automated unit tests and a small number of hardware boards running automated integration and system tests. Call it a *virtual farm* and a *board garden* to lean into the visual metaphor even more:

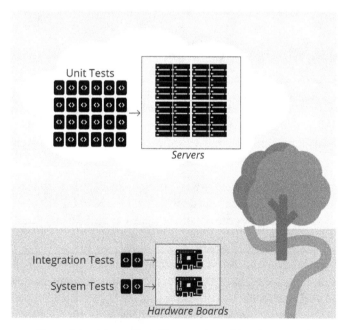

Figure 9.1 – A large virtual farm and small hardware garden

Examples of CI flows

Now we will look at implementing three different types of CI flows, ranging from very simple to more advanced. As referenced in *Chapter 3*, *Selecting the Right Tools*, automated testing environments have a natural trade-off: simpler implementations often lead to less test replicability and setup longevity, and vice versa. The goal of showing these three examples is to reduce the barrier to implementing a quality CI flow for your specific project's needs.

The three examples will be as follows:

- Simple test scripts
- In-house board farm setup
- A virtual farm with AVH

1 – Simple test scripts

To replicate the example in this section, you will be using the following:

Platform	NXP LPC55S69-EVK
Software	hello world
Environment	Personal Computer
Host OS	Windows
Compiler	Arm Compiler for Embedded
IDE	Keil MDK-Community

This example describes a flow useful for an individual developer looking to run a suite of builds and tests locally to a board sitting on their desk. The flow is not intended for large-scale projects but is a step up from no tests at all. It is lightweight and a good choice for quick prototypes or when starting a larger project.

The goal is to verify that after making some software changes, the code base still compiles and runs without error. We will use the *hello world*-provided software for the NXP LCP55S69-EVK board from the CMSIS-Pack in Keil MDK, which you can obtain the same way as the *blinky* example for the NXP board in *Chapter 4*, *Booting to Main*.

This Windows batch script has two commands. The first automatically builds the hello world project and sends the output to a file called BUILD_OUTPUT.txt, and the second automatically runs the hello world project and sends the output to a file called RUN_OUTPUT.txt. To put this in context, a developer could run this script after making interactive changes to their code and merging it with others to verify that nothing broke at a high level. It assumes you are in the same directory as the µVision project:

```
C:\Keil_v5\UV4\UV4.exe -r .\hello_world.uvprojx -o  BUILD_
OUTPUT.txt -j0
C:\Keil_v5\UV4\UV4.exe -f .\hello_world.uvprojx -o  RUN_OUTPUT.
txt    -j0
```

These commands call the µVision IDE but run it in a headless mode without the GUI due to the -j0 option. The -r option specifies a rebuild, while -f is for flashing the board. When running with the NXP board plugged in, we get an output similar to the following in BUILD_OUTPUT.txt:

```
*** Using Compiler 'V6.18', folder: 'C:\Keil_v5\ARM\ARMCLANG\
Bin'
Rebuild target 'hello_world debug'
compiling board.c...
...
compiling TransformFunctions.c...
linking...
Program Size: Code=6700 RO-data=1272 RW-data=2056 ZI-data=3172
"debug\hello_world.out" - 0 Error(s), 0 Warning(s).
Build Time Elapsed:  00:00:07
```

The following output is in RUN_OUTPUT.txt:

```
Load "debug\\hello_world.out"
Info: LPC55xx connect script start
Info: APIDR: 0x002A0000
Info: DPIDR: 0x6BA02477
Info: LPC55xx connect script end
Info: AP0 DIS: 0
Erase Done.Programming Done.Verify OK.Flash Load finished at
17:23:30
```

Checking manually, we can see the build was successful with no errors or warnings, and the flashing was verified and successfully flashed the board. As an example of an error state, if we run the same script without the board plugged in, RUN_OUTPUT.txt looks as follows:

```
Load "debug\\hello_world.out"
Internal DLL Error
```

```
Error: Flash Download failed  -  Target DLL has been cancelled
Flash Load finished at 17:19:42
```

This example does not explicitly test for any results, nor does it automatically read the output of these files to inform of an error, but both expansions are possible with more scripting. Implementing this simple local scripting is a low-overhead CI flow that really is just manual testing sped up. A proper CI flow is kicked off automatically on a developer code commit, which the next example will highlight.

2 – In-house board farm setup

To replicate the example in this section, you will be using the following:

Platform	Server
Software	Dot product
Environment	Jenkins
Host OS	Linux
Compiler	GCC (for x86)
IDE	-

This example describes a more robust CI flow, used at many companies today, of a central in-house server running a suite of regression tests on developer commits. Traditionally, this server is connected to a proper board farm, with dozens of boards connected and ready to be accessed. This example will feature the same setup but will swap the targets from boards to native host code testing, highlighting another style of regression testing in context.

The goal is to validate that after any developer commits to a certain code base, the software still builds and runs correctly, and gives the expected results. Upon committing to a specific code base, a Linux computer running the **Jenkins** CI controller will automatically detect this change and perform a build and run. To accomplish this, we need to initialize the tools in these steps:

1. Installing Jenkins
2. Publishing Jenkins
3. Creating a GitHub webhook
4. Creating a Jenkins pipeline

Installing Jenkins

The first step is to install Jenkins on an Ubuntu Linux machine. You can use an AWS or GCP instance if desired for this example – if you do so, then make sure to connect over VNC or forward the Jenkins port 8080 to configure Jenkins via a web browser. We are using a Linux laptop on our desk and we recommend the same if you plan to extend this example to connect to a physical hardware board later.

To install Jenkins, you can follow the instruction at this link to obtain the most recent Jenkins version using a few commands: `https://pkg.jenkins.io/debian-stable/`. After installation, navigate to your localhost port 8080 on any web browser: `localhost:8080/`. Follow the initial setup instructions. Unlock the software with a secret password in a file, select the **Install suggested plugins** button, and then set a username and password. This will fully install Jenkins, which will automatically start when restarting your computer, running as a service in the background.

Publishing Jenkins

We now need a way for a GitHub repository to notify Jenkins when a change has been made to trigger a `build/run` test script. In an active project in a company, there may be several differences in implementation here. The code base may be on a local Git repository connecting to an in-house CI controller machine, or a public GitHub repository connecting to AWS EC2 instances. This example will connect our Linux PC to a public GitHub repository over the internet, showcasing the functionality.

To do this, we will need to pipe our Jenkins service running on a local PC to a public GitHub repository. ngrok provides a simple way to accomplish this for testing purposes, as it is free for non-commercial use. First, download the ngrok package from this site: `https://ngrok.com/download`. Then, extract it from the terminal:

```
sudo tar xvzf ~/Downloads/ngrok-v3-stable-linux-amd64.tgz -C /
usr/local/bin
```

Add an `authtoken` (which you can get during signup, ensuring you have a unique token):

```
ngrok config add-authtoken <token>
```

Then, start a tunnel on port 8080 for Jenkins:

```
ngrok http 8080
```

This will create a tunnel for your Jenkins service to be accessible over the internet. You will see a popup as follows:

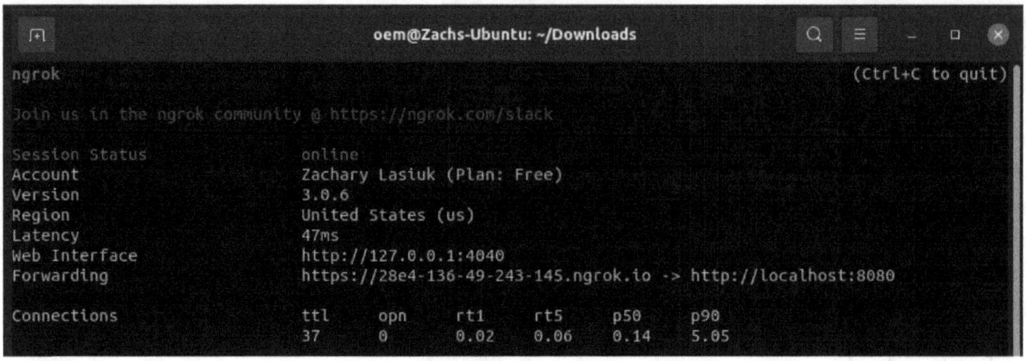

Figure 9.2 – The command line after starting ngrok

Navigating to the web address under `Forwarding` will connect to your Jenkins service. This is the URL we will provide to GitHub.

> **Important note**
>
> We only recommend keeping this open during your testing, not exposing your computer to the public internet for longer than trial prototyping. Proper security measures should be put in place in a production environment.

Creating a GitHub webhook

Next, we will create a webhook for a GitHub repository to notify Jenkins of any commit changes. To do this, see the following:

1. Fork the repository for this book into your own account.
2. Navigate to **Settings** | **General** | **Webhooks** to add a hook.
3. Under **Payload URL**, enter the URL from the previous step with `/github-webhook/` at the end. For this example, disable SSL verification (again, this is only recommended for brief prototyping).
4. Trigger the webhook on **Just the push event**, which will notify Jenkins of any commit change to the repository.

5. Finally, select **Update webhook** to finish:

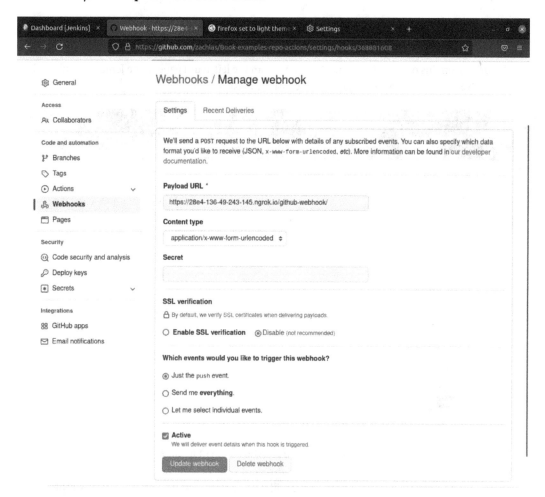

Figure 9.3 – Setting up a GitHub webhook

Your GitHub repository is now set to trigger a Jenkins job upon changes.

Creating a Jenkins pipeline

We will now tell Jenkins to listen to our GitHub repository in a specific testing pipeline and kick off some tests:

1.   Return to Jenkins on your Linux machine, and from the left-hand pane of the Jenkins dashboard, select **New Item**. Select **Freestyle project** from the list.

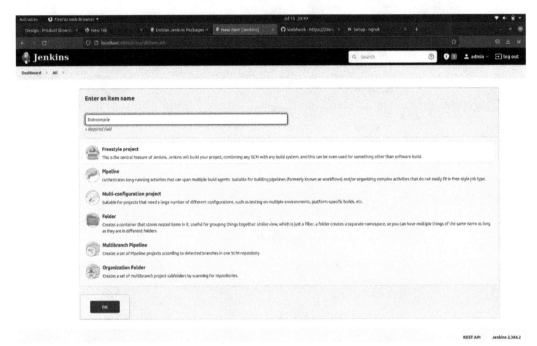

Figure 9.4 – Setting up a freestyle project in Jenkins

2.   Here, we name it `Dot-compile`, as we will be testing the build and running a simple dot product file used in a previous chapter.

3.   To set up the freestyle project correctly, provide any description first and head to the **Source Code Management** section. Here, you will need to provide a GitHub repository URL. To ensure a validated connection, you can leverage a GitHub personal access token. This token has replaced passwords for authenticating Git operations. To generate a new token, follow these instructions from the GitHub docs: `https://docs.github.com/en/authentication/keeping-your-account-and-data-secure/creating-a-personal-access-token`. Make sure to give your token full control of repository hooks for Jenkins to properly interact with GitHub.

4. After this process, you will want the repository URL to look as follows:

    ```
    https://<access token>@github.com/<userName>/<repository>.
    git
    ```

5. Next, under **Build Triggers**, check the checkbox labeled **GitHub hook trigger for GITScm polling**. This will trigger this job on every code push event.

6. Finally, under **Build**, we specify what commands Jenkins will run after being triggered. In this example, we give it two separate build events. The first is build, which navigates to the correct directory on the Linux machine where Jenkins automatically pulls the updated code, removes any previous build artifacts, and compiles a new dot-simple executable:

    ```
    echo "Starting build…"
    cd /var/lib/jenkins/workspace/Dot-compile/dot-simple
    rm -f dot-simple
    gcc dot-simple.c -o dot-simple
    echo "Build complete."
    ```

    The second event is really a run event that gives execution privileges to the generated file and runs it:

    ```
    echo "running app…"
    cd /var/lib/jenkins/workspace/Dot-compile/dot-simple
    chmod +x dot-simple
    ./dot-simple
    ```

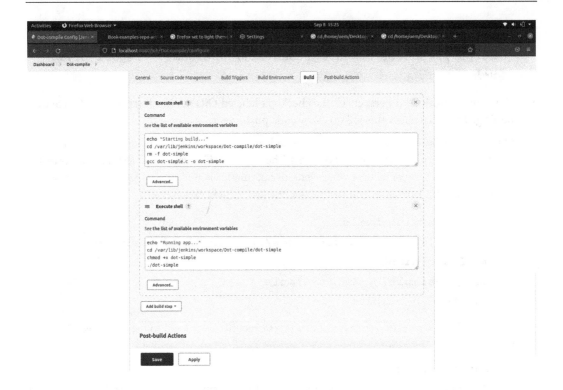

Figure 9.5 – Specifying the commands for the CI pipeline to execute

As mentioned previously, this example is running the dot-product algorithm natively on the Linux computer without the boot code specific to any Cortex-M hardware. These commands can be modified to build and run for the NXP board or Pico board.

7. Lastly, in **Post-Build Actions**, you can specify an email address to be notified on each job event, which is helpful for staying up to date with automatic runs.

8. Apply the changes and save the job to finish this setup.

You are now ready to test this out in practice.

Running your new CI pipeline

Now, try to push any change to the forked GitHub repository while keeping an eye on the bottom-left-hand side of your Jenkins Dot-compile job screen. Make sure you are viewing localhost:8080/job/Dot-compile/ to see the run happen in real time. After you commit to the GitHub repository, you should see a new build happening in real time. Upon clicking on that job, you should see the following output, indicating a successful build and run with the latest version of your code:

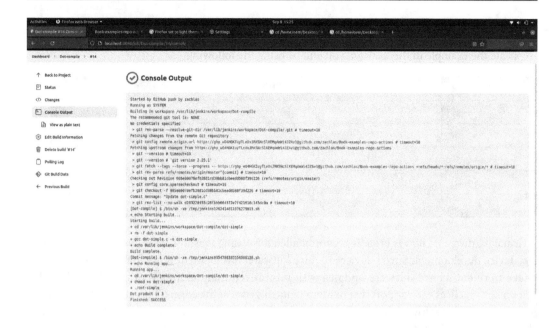

Figure 9.6 – The output of the CI pipeline

> **Important note**
>
> The newly updated GitHub repo is being downloaded to a specific location on the Linux PC to run builds and tests on. With this setup, you can make as many commits as you want to the repository. Other developers can as well – the same jobs will run every time.

While the overhead to set up this flow can be complex at first, the returning value from these tests can be excellent. Proper board farm setups run many tests in parallel on many connected boards, giving rapid feedback to developers if a code change has caused any issues and where those issues have arisen. CI flows can combine running tests on boards and the host machine – it depends on the specific needs of the software project. If running code on the host machine for proper development, a container system such as Docker is optimal to provide consistent results and system isolation.

Our recommendation for developers is to use the concepts of this board farm approach on a small scale for integration and system tests, in a small board garden. In parallel, scaling up unit tests on virtual platforms such as AVH can dramatically increase scalability. The next example will describe how to set up regression tests on AVH.

3 – A virtual farm with AVH

To replicate the example in this section, you will be using the following:

Platform	Arm Virtual Hardware – Cortex-M7
Software	AVH-AWS_MQTT_Demo
Environment	GitHub Actions
Host OS	Linux running on AWS AMI
Compiler	Arm Compiler for Embedded
IDE	-

The virtual hardware CI flow is broadly recommended for scaling up software testing, specifically for smart device development. Smart devices require software development flows that allow engineers to make frequent secure software updates with minimized manual intervention. Cloud-based development practices can support this new era of intelligence at the edge. CI practices allow for the robust automation of software development and allow data scientists to independently fine-tune and update ML models without needing any manual integration of firmware. Especially important for ML updates is the concept of **Continuous Delivery (CD)**, which extends CI from just validating software to deploying software automatically, keeping updated working code in production at all times.

With AVH, you can not only develop your software without having access to a board but can also run and scale the CI infrastructure in the cloud, with potentially several hundred virtual boards being launched in the cloud in seconds and all test suites running in parallel.

We will now look at how to set up a development workflow with cloud-based CI for testing an embedded application on AVH.

Understanding the GitHub Actions workflow

Similar to the previous example, a developer commits their code to a shared repository, which then triggers the execution of a CI pipeline. This example, instead of using Jenkins as the CI controller, uses **GitHub Actions** instead. This CI pipeline includes automated building and testing of the application on the AVH target that is running in an AWS AMI. It also includes posting the build and test results of your code to a GitHub Actions repository to make viewing the status simple:

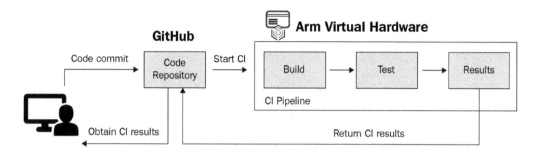

Figure 9.7 – A flow diagram of a GitHub Actions-enabled CI pipeline

In this example, we will leverage the AVH-adapted version of the AWS MQTT demo software project to detail the steps required for setting up a CI pipeline using GitHub Actions. The source code for this project is at `https://github.com/ARM-software/AVH-AWS_MQTT_Demo`. At a high level, this software project runs an IoT application on a Cortex-M7 AVH FVP and communicates to the AWS cloud services. While we are not going into the details of the application itself and rather focusing on steps to create a CI pipeline for it using GitHub Actions, you can read about the details of the application at `https://arm-software.github.io/AVH/main/examples/html/aws_mqtt.html` and `https://docs.aws.amazon.com/freertos/latest/userguide/mqtt-demo-ma.html`.

In contrast to the last section, where we initialized the Jenkins CI job in a GUI, GitHub Actions uses workflows that are defined in the `.github/workflows` directory of your repository to set up and execute the CI pipeline. These workflows are written in YAML and you can learn about them in complete detail at `https://docs.github.com/en/actions/using-workflows/about-workflows`.

The GitHub Actions workflow for this example can be found here: `https://github.com/ARM-software/AVH-AWS_MQTT_Demo/blob/main/.github/workflows/cortex_m_virtual_hardware.yml`.

We will walk through this workflow file and break down what each snippet of the workflow does. This is the first bit of code:

```
name: Cortex-M7 Arm Virtual Hardware

on:
  push:
  pull_request:
  workflow_dispatch:
```

Each workflow can be triggered by different events. The Cortex-M7 Arm Virtual workflow is triggered by either a `push` or `pull` request submitted into the code repository. This workflow can also be triggered manually either from the GitHub Actions GUI browser interface or using the API:

```
env:
  AWS_ACCESS_KEY_ID: ${{ secrets.AWS_ACCESS_KEY_ID }}
  AWS_SECRET_ACCESS_KEY: ${{ secrets.AWS_SECRET_ACCESS_KEY }}
  AWS_DEFAULT_REGION: ${{ secrets.AWS_DEFAULT_REGION }}
  AWS_S3_BUCKET_NAME: ${{ secrets.AWS_S3_BUCKET_NAME }}
  AWS_IAM_PROFILE: ${{ secrets.AWS_IAM_PROFILE }}
  AWS_SECURITY_GROUP_ID: ${{ secrets.AWS_SECURITY_GROUP_ID }}
  AWS_SUBNET_ID: ${{ secrets.AWS_SUBNET_ID }}
```

As this application is running on the AVH target in the AWS AMI, all the required AWS-sensitive credentials needed to start a new AMI instance must be provided as GitHub secrets. For the workflow to access these GitHub secrets, the preceding environment variables are set. If you are not already familiar with GitHub secrets and how to work with them, this document from GitHub (`https://docs.github.com/en/actions/security-guides/encrypted-secrets`) introduces the concept well:

```
jobs:
  cortex_m_generic:
    runs-on: ubuntu-latest

    name: Cortex-M7 Virtual Hardware Target
    steps:
    - name: Checkout
      uses: actions/checkout@v2

    - name: Set up Python 3.10
      uses: actions/setup-python@v2
      with:
        python-version: '3.10'
```

GitHub Actions workflows require a runner that executes the job and all the steps defined here. You can use either a GitHub-hosted runner or provide your own self-hosted runner for the jobs. Here, we are using a GitHub-hosted Ubuntu runner, which enables us to go completely hardware free – no need for in-house servers or board farms here. Standard GitHub Actions are then used to check out this repository, making it available in the runner's workspace, and installing Python version 3:

```
- name: Install AVH Client for Python
  run: |
```

```
        pip install git+https://github.com/ARM-software/
avhclient.git@v0.1
```

In the next step, we install something called the AVH Client. The AVH Client is essentially a Python module and it is used here to manage the connection to the AMI, well as for uploading, building, and running the application on the AMI:

```
- name: Prepare test suite
      env:
        MQTT_BROKER_ENDPOINT: ${{ secrets.MQTT_BROKER_ENDPOINT
}}
        IOT_THING_NAME: ${{ secrets.IOT_THING_NAME }}
        CLIENT_CERTIFICATE_PEM: ${{ secrets.CLIENT_CERTIFICATE_
PEM }}
        CLIENT_PRIVATE_KEY_PEM: ${{ secrets.CLIENT_PRIVATE_KEY_
PEM }}
      run: |
        cd amazon-freertos/demos/include
        envsubst <aws_clientcredential.h.in >aws_
clientcredential.h
        envsubst <aws_clientcredential_keys.h.in >aws_
clientcredential_keys.h
```

There are certain connection and security parameters that need to be set for communication between the AWS IoT service and the application we run on the AMI. We get these connection parameters from the AWS IoT Core service that we described in the previous chapter. These parameters are saved as GitHub secrets for this repository's **Actions** – as with the AWS credentials – and set as environment variables to be accessed by the workflow. Once they are made accessible, these settings are passed to the application by substituting through the environment variables:

```
- name: Run tests
      id: avh
      run: |
        avhclient -b aws execute --specfile avh.yml
```

Next, we run the AVH Client that we installed in the previous step. The AVH Client issues commands that get executed on the AVH AMI and runs the application under test. These commands are defined in the avh.yml file. Let's inspect this file at https://github.com/ARM-software/ AVH-AWS_MQTT_Demo/blob/main/avh.yml and dive into what is being executed with the AWS backend on the AVH AMI.

A deeper inspection of AVH.yml

First, the AVH.yml file issues the commands to set up the workspace on the EC2 instance with all the required application files from this repository. Next, Python is installed on this instance and build. py is executed to build and run the application on the Cortex-M7 FVP system. build.py issues the following command to build the AWS MQTT AVH application executable first:

```
/usr/bin/bash -c cbuild.sh --quiet AWS_MQTT_MutualAuth.VHT_
MPS2_Cortex-M7.cprj
```

If the build step is successful, the image.axf application executable is created. Next, this image is run on the Cortex-M7 FVP simulation target shown by the following command:

```
/opt/VHT/VHT_MPS2_Cortex-M7 --stat --simlimit 850 -f vht_
config.txt Objects/image.axf
```

The running AWS MQTT AVH application establishes an authenticated connection to the AWS MQTT broker. The application subscribes to MQTT topics and publishes messages that can be viewed on the AWS IoT MQTT client.

Finally, in the last step of the workflow, the build and simulation console output results of running this application are archived and uploaded as artifacts to the repository's **Action**. The output is generated by the following commands:

```
- name: Archive results
    uses: actions/upload-artifact@v2
    with:
      name: results
      path: |
        aws_mqtt-cm7-*.zip
        console-out-*.log
      retention-days: 1
      if-no-files-found: error
  if: always()
```

The test results can optionally be published, as well as using another action to make the test results easier to view and debug failures. The artifacts can be inspected to debug any test result failures. The default behavior specifies that artifacts are saved for 90 days after a run but in our example, we store the artifacts for only a day to minimize storage space usage, as shown by retention-days: 1.

We went through a suite of steps as part of this CI pipeline to build, run, and upload test results from the AWS MQTT application example running on the M7-FVP in the AVH AMI. The following is a snapshot from the **Actions** tab of our GitHub repository, listing all of the steps in a job that were automatically taken as part of this workflow:

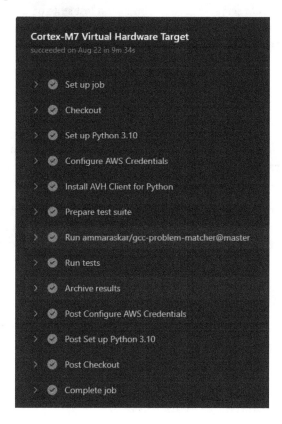

Figure 9.8 – The Actions tab of a successful GitHub Actions CI pipeline job

To summarize, by leveraging virtual hardware simulation models and applications running in the cloud and integrating them with robust CI pipelines, as in the one we outlined here, it has become much more feasible for software developers to easily test and validate their applications.

Summary

This chapter provided a broad set of resources for software developers looking to implement CI in Cortex-M-embedded projects. We discussed the values of utilizing automated testing first, noting how they outweigh the temporary drawbacks in terms of setup time. Then, we investigated how embedded developers, targeting custom hardware devices, have added constraints when setting up a CI pipeline in contrast with traditional website developers.

The last part of the chapter detailed three ways to implement automated testing with the identified constraints in mind. We started with a simple example suitable for individual and small projects: a straightforward script that automatically runs some builds and tests when executed. The second example represented the more traditional approach embedded software teams implement: large board farms connected to an on-prem server that runs a build and test suite on a code commit. The final example showcased the cutting-edge, scalable approach for effective embedded CI pipelines: large virtual farms for most tests with a small board garden for the few system-wide tests.

Make sure to select the level of CI complexity that makes sense for your code base and team size and you will avoid over-engineering or under-engineering a test flow. Whether you are on a small project personally creating your CI pipeline or a large team with a dedicated DevOps role, implementing a CI pipeline will increase your code quality and reduce development headaches.

There are many nuances in implementing your own CI pipeline, with plenty of opportunity for customization. The last chapter of this book will discuss the concept of code coverage, which adds significant visibility into code quality for all project stakeholders in real time.

The last chapter differs from the format of the other chapters in this book. It is split largely into two parts. It summarizes and adds assorted tips to the material presented thus far and then casts a view into the future to predict which embedded development skills will become more valuable in the years to come.

Further reading

- Integrating CI with Jenkins and Docker (note that the third part of the series does not exist and is effectively this book):

  - `https://community.arm.com/arm-community-blogs/b/tools-software-ides-blog/posts/implementing-embedded-continuous-integration-with-jenkins-and-docker-part-1`

  - `https://community.arm.com/arm-community-blogs/b/tools-software-ides-blog/posts/implementing-embedded-continuous-integration-with-jenkins-and-docker-part-2`

- The measurable value of using virtual farms over board farms: `https://community.arm.com/arm-community-blogs/b/tools-software-ides-blog/posts/slash-unit-testing-time-in-embedded-software-regression-testing`

- A comparison of unit, integration, and system tests: `https://u-tor.com/topic/system-vs-integration`

- An overview of AVH: `https://www.arm.com/products/development-tools/simulation/virtual-hardware`

- An overview of CD for embedded programmers (an old blog but a helpful introduction): `https://www.cloudbees.com/blog/5-aspects-makes-continuous-delivery-embedded-different`

- A detailed look at CD for ML models, known as MLOps: `https://cloud.google.com/architecture/mlops-continuous-delivery-and-automation-pipelines-in-machine-learning`

10
Looking Ahead

The world is constantly moving forward, and the devices and development techniques that support our culture move forward with it. This book has covered the most critical subjects to develop quality Cortex-M software in late 2022. Technologies that enable smart, connected, and secure devices are increasing in demand as individuals, companies, and governments see the value such products bring. Accordingly, we as developers must refine our development processes to create devices at a faster pace, lower cost, and higher quality every year.

The desire for smart, connected, and secure devices is as widespread as it is new. Twenty years ago, the concept of the **Internet of Things (IoT)** existed in name only. Devices were not connected to the internet by default, security was largely an afterthought (outside of sensitive industries), and machine learning was implemented only in cutting-edge applications. In twenty more years, the embedded software development landscape may be equally different.

This chapter is different from the previous chapters. It is intended to zoom out and provide general tips to be a successful Cortex-M software developer now and into the future. There are two distinct sections. The first contains tips and resources to develop quality Cortex-M devices today. The second section provides a look into the future. We look at over-arching trends and map how Cortex-M devices—and development techniques—will consequently evolve.

The now – tips to being a great developer

In each focus area of this book, we have attempted to atomize the topic, explain the underlying details, then showcase examples in a realistic context. In every area, there is so much to potentially cover that we selected the most salient points to write about and provide examples for. Two topics in particular (cloud development and testing/CI) are largely unutilized in embedded development and offer many avenues to increase your overall development efficiency. We will now expand on these two topics, presenting more options, context, and advice to move from being a *good* developer to a *great* one.

The cloud

As shown in *Chapter 8, Streamlining with the Cloud*, there are two main ways to think about leveraging the cloud: as a means to increase development efficiency, and as a means to increase device functionality and management when deployed. This section will offer further advice on both these axes.

Tip #1 – Try cloud service free tiers

Due to the fierce competition between cloud services and the fact that once people like a cloud provider, they largely stick with them, they all offer free tiers to evaluate their options without paying upfront. Google Cloud, Amazon Web Services, Oracle Cloud, Microsoft Azure, Alibaba Cloud, and IBM Cloud all enable you to try various services at no charge.

We highly recommend you try the free tier of multiple cloud service providers to learn about what cloud services are available. You might discover a service that would be prohibitively expensive for your team to create but is already offered. It will also introduce you to different techniques that can enable you to select what works best for your project. It is always helpful to get more context, and learning more in this area can pay off down the road.

Tip #2 – Saving and sharing AMIs

We have used **Amazon Machine Instances** (**AMIs**) throughout this book, specifically leveraging the Arm Virtual Hardware AMI to run examples on a virtual model of the Corstone-300 platform with the Arm Cortex-M55. We also discussed, toward the end of *Chapter 8*, how to develop code on EC2 instances with a VS Code IDE.

We suggest that you create your own custom EC2 instance for cloud-based development with all the tools and software you need pre-installed and save it as an AMI. You can then reuse it, spinning up new EC2 instances based on your custom AMI in a matter of seconds, to start with a fresh machine and all your required materials ready to go.

This is especially useful in projects that can leverage Arm Virtual Hardware during development, enabling you to code, compile, run, and debug all in the cloud with your favorite tools. You can then access this from anywhere without needing to bring a board with you when working from home or the office (or the beach). It is also useful in projects that have a non-trivial number of required tools and software to be present during development, as the AMI essentially *checkpoints* a machine's state right when you get it set up just the way you want.

A further benefit of this approach is the ability to share this AMI with your software development team. All developers in the same project can use the same AMI. This reduces overhead in each person setting up their machine with the right prerequisites and ensures everyone has the same environment and can reproduce any software issues.

Another advantage is the ability to try new things on the machine without fear of breaking your personal computer. Most developers have accidentally caused a computer to permanently break due to some low-level system tampering. When using virtual machines in the cloud, there is no fear and it's easy to be up and running minutes later due to this saved AMI development approach! It's freeing to know you can try things and just throw away the computer if they don't work out as hoped.

Tip #3 – Investigate other remote access tools

When developing embedded software in the cloud, a common annoyance is the quality of the remote connection. A slow or jerky connection to graphical tools can make development a hassle. We presented a few options in *Chapter 8*, including accessing VS Code over a web browser and opening a VNC connection. We recommend you look at other options to best fit your use case requirements. There are several good tools that exist today; here is an introduction to a few we have used with success:

- **Remote.It**: Offers a secure method to connect to multiple devices without exposing them to the public internet. Devices in this case can be edge IoT devices, and also cloud server instances.

- **Jfrog Connect**: Provides a secure way to manage IoT device fleets. Updating, controlling, monitoring, and managing are all use cases for Jfrog Connect.

> **Important note**
> Both Remote.It and Jfrog Connect can connect to Cortex-M-based devices and high-end cloud servers if desired.

- **NoMachine**: A like-for-like tool replacement for a VNC connection, offering a graphical connection to a remote machine that is exceptionally fast. It is not intended to manage IoT devices, but instead, to enable easier connection to a single cloud server to use it like a desktop. We have used this tool for personal, non-commercial projects to great success as it delivers a full remote connection to a remote server without any noticeable lag. Note that it does require a license for commercial applications.

Tip #4 – Cloud connector services

There are numerous cloud services available that connect IoT devices and collect IoT data. In *Chapter 9, Implementing Continuous Integration*, we used an example demonstrating how to connect an IoT device to AWS services and send messages using **Message Queueing Telemetry Transport (MQTT)**. MQTT is a simple messaging protocol for low bandwidth connections and constrained devices such as Cortex-M IoT devices. The example software demonstrated how to set up an AWS *thing*, a representation of a physical device or sensor.

All major cloud service providers have IoT services to connect devices and collect data. There are numerous examples available to learn about the device connection process, data transmission, and ways to visualize the collected data. Beyond cloud service providers, there are several companies offering dedicated IoT products for various markets. Additional services include device management, provisioning new devices, remote software update functionality, and more. Some services focus on connectivity when Wi-Fi is not possible using cellular solutions and LoRa and LoRaWAN protocols for long-distance radio transmission.

Even though it's not based on Cortex-M, a good way to learn about cloud services for IoT is to get a Raspberry Pi 3, Pi 4, or Pi Zero 2 W to learn. These devices run Linux on Cortex-A, but having Linux makes it very easy to learn the basics of IoT. There are many tutorials available, and the concepts translate well to Cortex-M IoT projects. One good place to start is Balena (`https://www.balena.io/`).

If Linux is not your thing, look for tutorials with the Raspberry Pi Pico W, and try out the numerous tutorials available. Collecting temperature data from a sensor and sending it to a cloud service is an easy place to start.

One of my favorite services is Initial State (`https://www.initialstate.com/`). It provides one of the easiest platforms to write simple code and stream to a data bucket. You can view the collected data with ease, and custom dashboards can be created in minutes. For some project ideas, visit `https://www.initialstate.com/learn/pi/`.

One of the trends in 2022 is Matter. Matter is an open source project that aims to connect devices from many vendors together in the "*smart home.*" All of us have smart home devices and are likely familiar with their successes and failures as consumers trying to get them to work. Compatibility is an industry challenge with room for improvement. To check out more, look at the GitHub project at `https://github.com/project-chip/connectedhomeip`. For a higher-level description of the Matter project, see the home page here: `https://buildwithmatter.com`.

Testing, CI, and safety

As shown in the previous chapter, there is a multitude of benefits when turning a traditional *board farm* into a *board garden with a virtual farm*. This section will cover separate but related topics around testing best practices that can add value to embedded CI flows, primarily around code coverage concepts and test implementation tips.

Basics of code coverage

Chapter 9 mentioned code coverage only briefly in passing, as a technique to avoid production errors when using CI. Code coverage, perhaps more accurately called test coverage, measures the proportion of your code exercised by an automated test suite. It can range from 0% (meaning no part of your code is being tested) to 100% (meaning every line of your code is being tested).

Having a higher code coverage leads to higher quality code, as you and your team can see the *health* of your code base in real time. Measuring your code coverage can be added as the last step to your CI pipeline while the CI manager, such as Jenkins and GitHub Actions, can be configured to automatically display the code coverage percentage. Modern code coverage tools will even identify which files are covered well and which are not exercised, indicating where to focus your attention to increase coverage. Without measuring code coverage, you don't know whether your code base is working as intended or whether there are bugs waiting to be exposed in production when your software is used in unexpected ways.

If you investigate adding code coverage to your project, note that the Cobertura format (Spanish for *coverage*) is the most popular. It was intended for Java-based code but can be utilized by embedded developers in other formats by using a compatible testing library that outputs in the same format. It is open source and is supported by the most popular managers such as Jenkins and GitHub Actions. There are other options that are commercially available and combine a test framework and code coverage measurement for embedded software, such as Bullseye Coverage and VectorCAST.

The downside of implementing code coverage is time. It takes time to write tests that cover every line of your code. There are a few guidelines that can help your project get the most value from having code coverage without overdoing it:

For most projects, 75% code coverage is the inflection point where achieving more coverage becomes unrealistic and the cost outweighs the bug-catching reward.

Note that 100% code coverage does not mean you will identify 100% of faults, errors, and bugs; it is estimated it will expose about 50% of potential issues by the nature of how tests are being performed.

Don't choose a code coverage goal before measuring your existing baseline for what your code coverage is today. Overestimation is common.

Your specific coverage goals will depend on your project's resources, design testability, and cost of failure in production.

With this overview of code coverage value and limitations, we will now discuss the different types of code coverage available.

Types of code coverage

We mentioned that code coverage measures the proportion of your code *exercised* by an automated test suite. But what does *exercised* really mean? This concept is different than splitting up tests by unit, integration, and system tests, which segment tests by scope. Unit tests look at small units of your code base, whereas system tests look at your whole system.

There are many different ways to measure how thoroughly tests *exercise* your code. The most obvious solution is *line coverage*. It essentially measures the number of lines that were tested, the total number of lines in your software, translated into a percentage. Line coverage is the oldest and perhaps more intuitive method of measuring code coverage, but, as is often the case, the first idea you think of is probably not the best solution. It can give unhelpful numbers when measuring common code situations. For example, suppose an `if-else` statement contains 1 statement in the `if` clause and 99 statements in the `else` clause. As a test exercises one of the two possible paths, statement coverage gives extreme results: either 1% or 99% coverage. This is not a helpful way to think about measuring code health and the percentage of code base coverage but is still a useful tool in the toolbox.

Note that there are often alternate names for the same measurement; line coverage is also referred to as statement coverage and C0 code coverage. The C0 references a quick way to talk about the three simplest types of code coverage. C1 is *branch coverage*, which measures whether each control structure (such as an `if-else` statement or `while` statement) has been evaluated to be both `True` and `False`. C2 is *condition coverage,* which measures whether each Boolean variable in a conditional statement has been tested in all combinations of `True` and `False`.

The following table summarizes the most common types of code coverage for your reference. We encourage you to research examples of these different types in practice as that is the easiest way to understand what exactly they are measuring. Simply searching the metric name in Google is sufficient:

Name	Definition	Benefits	Drawbacks
Statement Coverage (Line coverage)	Reports if each executable statement is encountered	Discover control flow issues	Insensitive to some control structures; reports if loop body was reached, not if terminated
Branch Coverage (Decision coverage)	Reports whether Boolean in control statements evaluates to `True` or `False`	Simplicity	Insensitive to "short-circuit" operators; insensitive to compiler optimizations
Condition Coverage	Reports the `True` or `False` outcome of each condition (an operand of a legal operator)	Thorough test	Full condition coverage does not guarantee full decision coverage
`Modified Condition / Decision Coverage` (`MC/DC`)	Every entry/exit point is hit, every decision has had all possible outcomes, all possible directions taken, and each condition has independently affected a decision's outcome, all at least once	One of the most thorough coverage types	Insensitive to short-circuit operators
Call Coverage (Function coverage)	Reports if each function call is executed	Useful in verifying component interaction	Limited scope being analyzed, good as an additional metric to measure

Table 10.1 – Summary of basic code coverage types

You can choose to implement any of these code coverage metrics, depending on the needs of your code base. There are many more code coverage types that are less common that may be interesting to explore, such as Object Code Branch Coverage, Loop Coverage, Race Coverage, Relational Operator Coverage, and Table Coverage. The area of code coverage, measuring how well your tests exercise your code base, is still only one (albeit the most common) way to measure your code base health. With a proper CI flow in place, you can measure any metrics that you want to track progress, developer efficiency, and more. Here are some examples for you to think about:

- Lines/function
- Tests/function
- Number of hits/function
- Number of hits/code line
- Code change/time
- Different developer edits/function
- Number of GitHub check-ins/developer
- Number of failures/GitHub check-in

You should now have a baseline understanding of what code coverage metrics there are. The next section will cover the different ways to actually implement tests and code coverage in your test suite.

Implementing tests and code coverage

There are several ways to implement tests and code coverage in your project. This section will give an overview of some options that we have encountered, and subsequent considerations.

VectorCAST, Bullseye, and Parasoft are all excellent tools to instrument your embedded code base with tests. Implementing your tests with these tools enables you to automatically measure various code coverage metrics with their built-in capabilities. They are commercially available and are commonly used in the Cortex-M software space to create test suites for medium to large projects.

For small projects, there are several test frameworks that offer no or minimal code coverage metrics but are straightforward to implement. For tests that you plan to run on a host machine and not your embedded system, GoogleTest and CppUTest are two simple options. To mock and stub certain functions or hardware peripherals that you want to replicate in a testing environment, the Fake Function Framework is a very simple offering, as is the Ceedling tool. You can investigate any of these options if interested in learning more, as I (Zach) think this topic is best understood through exploration and experimentation. With these options, you would need to integrate a separate code coverage measurement technique/tool if desired.

If you are using the GCC compiler, the Gcov tool adds the ability to measure line coverage without the need to modify your existing source code or tests. It is a standard tool with the GNU Compiler suite. Simply call gcc from the command line with extra flags, as shown here:

```
$ gcc -fprofile-arcs -ftest-coverage test.c
```

This will generate an instrumented executable file as well as some profile files. You can then run gcov on the source file to generate a simple code coverage report:

```
$ gcov test.c
```

> **Important note**
> gcov only measures line coverage, and more complex code coverage measurements require more robust tools.

The embedded software testing space can be complex to navigate. For larger projects and teams, there are excellent commercial tools for developing test suites effectively, and should be leveraged appropriately. For smaller groups, the space is quite fragmented, and you should select a framework that you are familiar with and can achieve your goals with minimal overhead.

Now, let's transition away from discussing code coverage and return to a broader embedded software development theme. Examples are often the best way to learn what is possible and how to develop software better. The next section will provide references to project examples, useful code bases, and tools to improve your overall Cortex-M awareness.

Exploring useful examples and code

The Arm Cortex-M space moves fast. New boards and new example software are created every month, and it is helpful to any Cortex-M developer to be aware of emerging resources that can help your own development. These resources can be anything from new hardware, code bases, tools, or excellent examples showcasing what is possible today. Let's dive in!

Examples on the Raspberry Pi Pico

The Raspberry Pi Pico was released in 2021 on the RP2040, the first SoC designed by Raspberry Pi themselves. It has exploded in popularity, with the Arm Cortex-M community creating examples exploring what is possible on the Pico. Here are some examples that showcase Pico's possibilities.

Multi-core Pico

This example highlights a core (pun intended) feature of the RP2040 SoC on the Pico: two Cortex-M0+ processors. As discussed in *Chapter 1, Selecting the Right Hardware*, Arm optimized the Cortex-M0+ to minimize power and area. This comes at a cost of performance, making it difficult to run concurrent tasks on one Cortex-M0+ such as communication and sensing data.

The multi-core nature of the Pico offsets this disadvantage by enabling two tasks to run in parallel, one on each Cortex-M0+. This example highlights, in simple yet powerful terms, how to utilize this functionality with provided source code. You can view and replicate the example here:

```
https://learnembeddedsystems.co.uk/basic-multicore-pico-example
```

Multilingual Blinky

This example is a good reference to explore the different languages that the Pico can run. It shows how to go from *zero to blink* for several languages supported by the Pico at the time of publication. The languages included are as follows:

- C (naturally)
- MicroPython
- CircuitPython
- JavaScript
- Arduino
- Rust
- Lua
- Go
- FreeRTOS (an OS port, not a language port)

View the blog to implement these languages yourself:

```
https://www.raspberrypi.com/news/multilingual-blink-for-raspberry-pi-pico/
```

USB microphone

This example illustrates how flexible the RP2040 SoC can be, defining how to create a low-cost microphone cleverly named the *Mico*. It uses a software microphone library for the Pico and a small PDM microphone, as well as a custom-defined PCB to situate the components. You can view how to create your own custom Mico or learn from the software libraries provided for future projects. An example can be found here:

```
https://www.cnx-software.com/2021/12/31/mico-a-usb-microphone-based-on-raspberry-pi-rp2040-mcu/
```

Micro-ROS

This example showcases how to bring the Pico into the world of robotics, through a modified version of the **Robot Operating System (ROS)**. ROS (https://www.ros.org/) is a purpose-built suite of software, libraries, and tools to enable the creation of robotics applications. It is intended to be run on rich OSes such as Windows and Linux, targeting powerful computers controlling robot components. Historically, there has been a gap between easily controlling and communicating with resource-constrained microcontrollers in robotics, tasked with real-time behavior and reducing power draw.

The Micro-ROS project bridges the gap between large processors and tiny microcontrollers in robotic applications. If you are interested in exploring this project and its possibilities, you can find its overview here: `https://micro.ros.org/`. This blog on running Micro-ROS on the Pico is a great example of how to use the Pico in unique ways: `https://ubuntu.com/blog/getting-started-with-micro-ros-on-raspberry-pi-pico`.

Examples for machine learning

Going beyond examples for the Raspberry Pi Pico, there is more content on websites, forms, and tools that highlight various Cortex-M-based board capabilities. The following are some helpful examples in the machine learning space.

Image classification

This example walks you through how to create an edge device with the ability to recognize objects in your house. It is flexible, with several hardware board options that can use the same flow, including the Raspberry Pi 4. This tutorial focuses on simplifying the machine learning implementation with the Edge Impulse tool. Consider it a great resource to understand the end-to-end machine learning implementation process, including data collection, preprocessing, ML algorithm selection, model training, model validation, and flashing to your board. Find examples on the Edge Impulse website: `https://docs.edgeimpulse.com/docs/tutorials/image-classification`.

Speech recognition

This example enables you to port a TensorFlow Lite model to an edge device to perform on-device wake-word detection. It features the NXP i.MX RT1010 board based around a Cortex-M7. You can get hands-on experience with using the TensorFlow Lite Micro library and can use these same principles in other projects with related goals. See the example here: `https://www.hackster.io/naveenbskumar/speech-recognition-at-the-edge-40ba12`.

Code bases, tools, and other resources to leverage

Examples are useful resources when developing software. There are also helpful code bases to expedite software development, assorted tools to make development easier, and educational resources to learn more about a specific topic. Hundreds of these libraries, tools, and resources exist; there are too many to list here. Instead, we will focus on a select few that we see as adding substantial value to common themes in Cortex-M projects.

Pigweed

While still new and under development, Pigweed is poised to represent a step-change in usability for embedded development. Pigweed aims to enable faster and more reliable development to microcontrollers, bringing common advantages that web developers have (such as saving a file then instantly seeing the change on a browser, and reproducible environment initialization) into embedded development.

It was announced in 2020 headed by the Google Open Source program and is slowly growing with support from the embedded community. It already offers modules that simplify environment setup, development, code validation, unit testing, and even memory allocation. The Raspberry Pi Pico is under development as a target for Pigweed, and you can view the announcement and documentation here:

- `https://opensource.googleblog.com/2020/03/pigweed-collection-of-embedded-libraries.html`

- `https://pigweed.dev/`

DSP Education Kit

Digital Signal Processing (**DSP**) is a wide topic that can be quite complex to implement efficiently in certain use cases. Arm University, a program from Arm that provides resources to schools to educate people on the latest technology from Arm and its ecosystem, has an education kit dedicated to DSP. Publicly available on GitHub, the DSP Education Kit contains individual modules that focus on creating audio applications on Arm processors. It covers convolution, **fast Fourier transform** (**FFTs**), **finite impulse response** (**FIR**) filters, noise cancelation, predictive algorithms, and adaptive FIR filters, to name a few.

The GitHub repository offers example code with associated PowerPoint/Word documents that provide context. It is an excellent resource to learn audio algorithm implementation and utilizes an STM32F746G Discovery board. The repository is located here: `https://github.com/arm-university/Digital-Signal-Processing-Education-Kit`.

CMSIS tools

CMSIS libraries offer a simplification for Cortex-M software developers, and this book has shown several examples of leveraging CMSIS software. For more than a decade, CMSIS has provided software reuse across the Cortex-M microcontroller industry. It is a valuable learning tool and saves significant time in any project.

Developers continue to consume CMSIS in the form of CMSIS-Packs. A Pack is a delivery mechanism to package source code, libraries, documentation, and examples into an easy-to-consume bundle. Packs have always been easy to find and import into Cortex-M development tools such as Keil **Microcontroller Development Kit** (**MDK**) and other similar tools.

The Open-CMSIS-Pack project was recently started to open the development of the infrastructure and tools for Packs. This will make the standards and associated tools open source.

Open-CMSIS-Pack is important because, as we have highlighted, Cortex-M developers are taking advantage of new tools and technology using the cloud and automation. Historically, most Cortex-M development has been done on Windows PCs but is now shifting to the cloud and using other operating systems such as Linux and macOS. There is also increasing use of VS Code for microcontroller development. The Open-CMSIS-Pack project delivers new tools that can work on a variety of machines including Linux and macOS, and even machines that use the Arm architecture. This freedom and flexibility ensure that developers can continue to get the benefits of software reuse and take advantage of example code, no matter what kind of computer they use. If you have been using Packs with a traditional IDE, now is a good time to investigate the Open-CMSIS-Pack development tools.

View the repository here: `https://github.com/Open-CMSIS-Pack/devtools`.

Vectorization on the Cortex-M55 and Cortex-M85

The two newest Cortex-M processors both bring the power of vector processing to the Cortex-M family. Due to its recency, many microcontroller developers are not used to working with vectors in this low-resource context.

There are many available resources to learn more about this topic, summarized brilliantly on an ongoing basis by Joseph Yiu in the Arm Community. Here is the link to that page: `https://community.arm.com/arm-community-blogs/b/architectures-and-processors-blog/posts/armv8_2d00_m-based-processor-software-development-hints-and-tips`.

Reviewing these resources will give you a broad understanding of Cortex-M possibilities, and great places to start from today. In addition to these great community resources, there are several Arm projects being developed today (mid-2022) that will offer great value to Cortex-M software developers as they mature. Next is a brief overview of these projects to be aware of.

Official Arm projects under development

When we say *official* Arm projects, we mean projects that are managed and maintained by Arm employees under the Arm company umbrella. The primary place to look for these software projects is on GitHub under the Arm Software organization: `https://github.com/ARM-software/`. We have referred to several well-established repositories from this organization in this book already. Here are a few projects that Arm is actively developing to be helpful resources as they expand wider.

IoT SDK

The Arm Open-IoT SDK is intended to guide developers to explore and evaluate Arm IP, CMSIS APIs, and tools. Today, it is mainly focused on Arm Total Solution applications, with resources highlighting Arm Virtual Hardware, continuous integration, and the ML Embedded Evaluation Kit (all of which we have explored in this book). It also will include reference implementations for PSA (such as TF-M, covered in *Chapter 7*, *Enforcing Security*) and the Open-CMSIS-CDI, which is covered next. This is the GitHub link: `https://github.com/ARM-software/open-iot-sdk`.

Open-CMSIS-CDI

This project is new and launched in the summer of 2022. It is a collaboration between Arm and Linaro to define a common device interface for microcontrollers in the IoT. An ambitious goal is just getting started, planning on defining a common set of interfaces that will enable cloud service-to-device interaction and enable common IoT software stacks to run across Cortex-M-based devices with minimal porting required.

At the time of publication, there is very little public code on GitHub, but a repository has been established with the potential to fundamentally change how developers work on IoT projects. Keep an eye on this repository in the future: `https://github.com/Open-CMSIS-Pack/open-cmsis-cdi-spec`.

2D graphics library

The seemingly natural arc of technology is to start as new and difficult to use, then expand to mass-market adoption by being easy to use. This is primarily how computers became widespread: improving on the command-line-based interaction from older computers to a graphical interface that was easier for most people to work with. The IoT is at a similar point now, exacerbated by people's expectations due to their slick smartphone interfaces. IoT devices will require GUIs to be widely adopted.

Offering this from Cortex-M-based microcontrollers, running at tens or hundreds of MHz, can prove challenging. Most GUI libraries are built on top of rich OS drivers, enabling Linux developers but not Cortex-M developers. The Arm-2D project aims to solve this problem by accelerating fundamental low-level 2D image processing, abstracting away complexity to jump-start IoT GUI development. It is built on CMSIS components and is under active development. You can view the GitHub project at this link, which provides excellent context and examples to get started: `https://github.com/ARM-software/Arm-2D`.

So far in this chapter, we have reviewed examples, tools, software libraries, and Arm projects that you can reference to create better Cortex-M software. From here onward, we will focus not on what is possible now, but instead on what may be possible in the future.

The future – how trends will affect Cortex-M developers

In the final section of this book, we attempt to translate emerging societal and technological trends into their impact on Cortex-M software development. This is a very subjective undertaking, as predicting the future is fraught with biases and faulty assumptions, but we nonetheless will try.

We are at a critical inflection point in human history, with technological capabilities increasing exponentially as advancements in computation produce faster advancements of computation. Our society is rapidly reconfiguring around technology but is largely still based on centuries-old institutions; our governments were created when the newspaper was the primary mass communication method. Our human brains have largely remained the same for thousands of years, with our core instincts and intuition developing in a completely different society based around hunting and gathering. So, humanity now has a problem: we have Paleolithic emotions, medieval institutions, and godlike technology.

This fundamental clash is causing tectonic societal shifts and continual change that we can all feel in the early 2020s. There are three primary areas that we will focus on here, breaking down how these trends could influence the ways we as embedded software developers work:

- 5G and the Internet of Everything
- Environmental sustainability
- Decentralization of information

There are numerous other shifts occurring as well, such as job automation, which we do not have the space to cover in this book. We suggest that you investigate how societal shifts in your locale may alter how or what software you develop, as it can place you as knowledgeable in a critical area before others are aware of the need.

Trend 1 – 5G and the Internet of Everything

5G, for those not already aware, is the fifth generation of cellular mobile communications. It is, like its predecessors 4G, 3G, and 2G, a system that connects devices to the internet through cell towers. The 5G network represents a huge leap in capabilities over 4G and 4G LTE, enabling orders of magnitude better performance in traffic capacity, network efficiency, connection density, latency, and throughput.

We have not talked explicitly about 5G technology in this book as many readers will already be familiar with the topic. The topics we have discussed—such as security, machine learning, and cloud computing—will naturally become more essential as the 5G network expands. New IoT use cases that were previously prohibited by cost, complexity, or network capability will soon become possible (and profitable).

5G as an enabler

At the risk of sounding clichéd, a global 5G network will truly unlock the value of IoT. Opportunities will arise to make the world of "things" more effective and efficient. We will be able to better *understand* and *act* through "things" in the world around us. These "things" will be the next generation of IoT devices. I (Zach) like to break down these devices into two categories: sensors that help us understand, and actuators that help us act.

Sensing devices are already numerous and will grow to encompass a huge range of capabilities, measuring things such as air quality, traffic congestion, parking availability, trash levels, electricity consumption, drinking water quality, food freshness, ocean acidification levels, manufacturing defects, and so on. *Actuating devices* act on or in our world to achieve a goal, such as moving a car without human input, running factories, flying drones, providing dynamic city lighting, and thousands more.

While IoT is set to grow dramatically, its growth will still largely depend on the core software concepts we have discussed in this book. Concepts of particular importance include security, machine learning, and cloud development practices. We also expect that, due to 5G network capabilities, market opportunities, and societal priorities, IoT will grow broader and become the Internet of Everything. This means sensing and acting not only on limited elements of the world but on essential societal systems and even human beings.

The Internet of Everything

The term *Internet of Everything* is used to differentiate from the *Internet of Things* by the nature of what is being sensed and acted upon. The distinct terminology helps us think differently about what is possible in this solution space. The ability to automatically understand and modify the behaviors of people is fundamentally different than sensing and acting on inanimate objects. Even controlling critical infrastructure is different in kind from controlling other inanimate objects as they are essential to a functioning modern society.

Using the term *IoT* to describe what exists today versus the next generation of connected devices is like calling the first cars *mechanical horses*. Cars, in their forms and abilities, are so much more than horses that using the name *mechanical horses* limits our thinking about what they even do—and what they can affect.

The monitoring and modifying of the human body and behaviors is one hallmark of the Internet of Everything. Health wearables fall into this category, although today the most common devices, such as smart watches and fitness trackers, only monitor activity. Some specialized devices do act on our bodies today. Pacemakers stimulate heart muscles to beat and transmit status information to a smartphone. Continuous glucose monitors sense glucose in the bloodstream and dynamically allocate insulin as appropriate.

The next generation of *Internet of Everything* devices will take these specialized use cases to everyday human health. Imagine, while eating dinner, receiving a notification from your phone that cancerous cells were detected on your stomach lining. Luckily, you are told that your wristband has already injected nanobots into your bloodstream that have isolated and neutralized the threat. Medication will arrive at your door tomorrow that you can take if you experience any minor side effects, such as stomach discomfort.

This scenario may seem far-fetched and perhaps a bit sinister, but there are many possible ways to automatically monitor and modify someone's health that are within technological reach today. The natural human desire to be healthy, combined with the scale of the number of humans on earth, will create large monetary incentives to develop this type of product.

The second major hallmark of the Internet of Everything is the ubiquity of control over critical infrastructure. This includes the electricity grid, water treatment facilities, oil and gas pipelines, nuclear weapons facilities, and internet services, among other essentials of modern life. Previously, these facilities were controlled by a select few individuals with private access to physical resources. The proliferation of digital connectivity granted governments/companies the ability to automatically monitor and control this infrastructure.

This level of automation creates cost-saving efficiencies and improved service reliability. It also creates a larger attack surface for hackers to disrupt critical systems. There have already been examples of this happening to countries and companies all over the world. Security will be even more important in the future as more of humanity's critical infrastructure becomes digitally connected.

The Internet of Everything requires reliability, security, and privacy principles above and beyond the standards set for IoT devices today. Software developers targeting Cortex-M devices can start sharpening skills now to plan for the expansion of IoT and the emergence of the Internet of Everything.

Here are some takeaways for embedded software developers:

- Learn how to effectively implement security and machine learning on edge devices

- Learn techniques to provision and manage a large number of IoT devices deployed in the field at once

- Learn the reliability and safety capabilities of Cortex-M processors, and how to develop safety-certified software, to work in the critical infrastructure and medical industries

- Learn about ethical design in healthcare, and how to ethically modify human behavior with informed consent

The next trend has broad implications across all of society, and specific implications for Cortex-M developers: environmental sustainability.

Trend 2 – environmental sustainability

A core reason why technology has advanced so quickly in modern times is due to our access to hyper-compressed forms of energy, releasing thousands of years of solar energy at a time through buried hydrocarbons. Oil and natural gas were created over millions of years and are considered non-renewable resources; they cannot be replaced in a human-scale time frame. While we are not on the cusp of using all oil and natural gas in the world, it will become more expensive over time to extract and refine them.

In the process of burning fossil fuels at societal scales, Earth's climate has begun to change. This is considered a scientific fact (`https://climate.nasa.gov/scientific-consensus/`). As the effects of climate change become starker, investing in fossil fuels becomes less viable. The head of BlackRock, the world's largest asset manager, with $10 trillion under management, sends out an annual letter to CEOs that he invests in. His letter in 2022 indicated that BlackRock is *"asking companies to set short-, medium-, and long-term targets for greenhouse gas reductions."* This is an example of a large economic priority shift from fossil fuels to renewable energy sources, and pressure to decrease energy consumption generally.

Renewable energy challenges

Even if the world switches 100% to renewable energy, renewable energy sources such as wind, solar, and hydroelectric power cannot exactly replace the energy output of fossil fuels. Transporting a barrel of oil is straightforward enough; it can be done en masse to move energy from where it originated to where it is needed in a stable liquid form, storable, and ready to be burned by combustion engines when needed.

Transporting solar energy is much more complicated. Transmitting the energy requires high-voltage power lines that must be built at high up-front costs from new solar farm locations to local substations. Old power lines must be upgraded to handle the larger energy throughput. Solar energy also cannot be easily stored on location in a ready-to-use form. Furthermore, solar energy (and renewable energy in general) is also significantly less dense than non-renewable energy, requiring more energy to capture the same amount of power from other sources per unit volume.

All of this is to say that economic and moral pressure is being placed on maintaining and increasing our standard of living while drastically reducing the power required to do so. For developers in the Cortex-M space, this means focusing more on ultra-low power devices.

The ultra-low power solution

Ultra-low power refers to devices that require an exceptionally small amount of power to run, by today's standards. This enables devices based on Arm Cortex-M processors to be run by small batteries that last for many years or, in emerging cases, without batteries at all. Energy harvesting enables some devices to gather energy from the surrounding environment to power their components. Sources include indoor solar, ambient RF waves, vibration, and thermal gradients.

There are several emerging examples today of battery-less devices. The University of Michigan has spun out Everactive, a startup focused on making battery-less sensors: `https://everactive.com/`. RELOC has also developed a Bluetooth 5.0 module that can harvest energy for battery-less sensors. It is called the RM_BE1 and is based on a Renesas MCU featuring the Arm Cortex-M0+. Universal Electronics has announced an SoC that can harvest energy from ambient RF and indoor light, intended for voice-activated remote controls. It will also leverage Arm TrustZone security technology. That announcement can be found here: `https://www.uei.com/news/uei-unveils-extreme-low-power-chip-platform`.

Creating sustainable devices that are ultra-low power and last decades also addresses another looming problem: electronic waste and precious metal depletion. The currently known accessible supply of lithium (an essential element for batteries today) is smaller than the proposed rate of global electrification. As lithium demand outpaces accessible supply, the requirements for ultra-low power devices will likely skyrocket. Software developers targeting Cortex-M devices can start gaining skills now to plan for this likely future focus on ultra-low power consumption in IoT.

Here are some takeaways for embedded software developers:

- Learn how to implement Cortex-M power optimization techniques in software, such as hibernation states

- Learn about how ultra-low power applications restrict software capabilities, and how to work within these constraints to deliver a functional device

- Learn about battery-less devices and energy harvesting as the technology advances over time

- Learn how to extend your device's life cycle to decades instead of years, including field updates and repairability

The next trend focuses on a broad trend with a few different repercussions for the technology industry: the decentralization of information.

Trend 3 – decentralization of information

When the internet started to become mainstream in the 1990s and early 2000s, it was anticipated to be a medium where individuals could interact directly with other individuals on their own terms. Radio had strict frequency allocation maps that restricted public communications, and television was a one-way medium communicating information from media monopolies to the public. The internet was supposed to be decentralized, with the power to create and share imbued to individuals.

The centralized web

Over the last two decades, it has become clear that monopolies naturally form on the internet due to network effects. A network effect occurs when the value of something increases the more people use it, leading to exponential value creation. Think of social media companies; it is easier to connect with friends when they are all on one social media network, and new social media companies cannot compete with existing ones that have all people on it. Why switch to a social media platform when there is no one to be social with?

This is as true for social media platforms as it is for search-based companies. The more people search on a given website, the more data they can collect and the better they can get at serving results people want. Advertising-based business models on the internet follow the same dynamics, resulting in a handful of hyper-scale companies that centrally control much of the internet landscape.

The most valuable asset these massive internet companies has is their users' personal information and data (including the machine learning models based on this data). The value of large quantities of personal information also leads these companies to sell user data to make money, through commercial and political advertising. In the past few years, public sentiment has started to turn against the free gathering and inappropriate sharing of personal data, exacerbated by certain scandals. First envisioned as a decentralized medium to empower individuals, the internet from the 2000s to today has largely done the opposite. There is still a societal desire to develop technology that redistributes power from centralized sources to individuals, however. This desire is currently dovetailing with new decentralization technologies, such as blockchain, and we are now on the cusp of a new type of internet: **Web 3.0**.

Web 3.0 and decentralized IoT

The term *Web 3.0* (also referred to as **Web3**) is largely a shorthand buzzword describing a larger idea: a new iteration of the internet that leverages decentralization and blockchain technologies. Blockchain technology is a key technology that enables mass decentralization. There are many resources explaining what blockchain is, what it can do, and its value. IBM has a succinct and helpful explanation here: `https://www.ibm.com/topics/what-is-blockchain`.

This new decentralized internet is just emerging and use cases are still being proven in practice. The potential range of Web3 applications is massive. Decentralized money, such as cryptocurrency, could fundamentally change how our economy operates. You will certainly be familiar with the popular example of Bitcoin, with hundreds of others being used today as well. Decentralized information on a blockchain has the potential to securely trace food through supply chains to minimize food waste. Decentralized data ownership could enable individuals to profit from their personal information instead of companies. The web browser Brave offers this today by enabling users to opt in to see targeted ads, who receive Basic Attention Tokens (a cryptocurrency) in exchange.

These possibilities will influence what skills embedded developers need to be successful in the Web3 space, with the largest areas being increased attention to security and data privacy. Another possibility is a shift in how IoT devices are managed to better align with the core value of Web3: decentralization.

Today, the predominant management model of IoT device networks is centralized. All IoT devices for a single use case are managed by a central node in the system. This arrangement makes it simple to push secure firmware updates to all devices and manage incoming device data, to take two examples. It also minimizes the amount of storage and processing power required on each device as most storage and processing is performed at a central server. Complying with data protection regulations, such as GDPR, becomes easier due to storing and managing information on a small number of well-defined server controllers.

At the opposite end of the spectrum is a distributed IoT network. Here, each device node acts completely autonomously and is interconnected with other network nodes. This requires more storage and processing capabilities to be built into each device, increasing costs. Issues also arise when trying to comply with data protection regulations, and maintenance requires more thought and effort. There are, however, many benefits to this appro   ach. A high potential to scale exists due to efficient task distribution; stability is formed from a lack of a single point of failure, and complex tasks can be processed quickly throughout the network nodes.

There is also the concept of a decentralized IoT network, which is between a centralized and distributed network in nature. Its IoT nodes are clustered into smaller networks that connect to one another through "super-nodes." The costs and benefits of this approach are between a centralized and distributed management model.

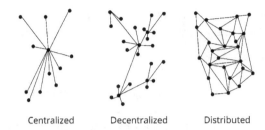

Centralized   Decentralized   Distributed

Figure 10.1 – Different network architectures, visualized

The overarching idea of moving power away from central sources is obeyed by both a distributed and decentralized IoT network. In practice, applying one of these management models to your next IoT project may require new ways of approaching things such as device management, firmware updates, secure storage locations, data privacy, and distributed computation.

In summary, as decentralized concepts grow more popular, embedded programmers should be ready to apply them. This means properly enabling the decentralized internet with IoT devices and creating decentralized/distributed IoT networks.

Here are some takeaways for embedded software developers:

- Learn about different business models involving people owning their own data, and how this will affect what is possible with ML at the edge

- Learn how to effectively implement security best practices on edge devices, with special attention to private user information

- Learn about the principles of distributed versus decentralized versus centralized IoT network models, and how to effectively manage IoT fleets without centralized control

To learn more about how to design technology products that take into account these (and more) societal trends, check out the Center for Humane Technology. They have excellent resources such as articles, podcasts, and expert interviews that expand on these kinds of topics (`https://www.humanetech.com/`).

The end

With that, our journey comes to an end. Throughout this book, we covered a lot of ground across various Cortex-M topics. *Part 1*, *Get Set Up*, detailed how to intelligently select the right hardware, software, and tools for your specific project requirements. *Part 2*, *Sharpen Your Skills*, showed how to implement key software topics (booting to main, optimizing performance, leveraging machine learning, and enforcing security) and development techniques (streamlining with the cloud and implementing continuous integration). This last chapter offered avenues to continue learning how to be a better Cortex-M software developer: helpful tips and examples to view today, projects to leverage in the coming months, and our subjective previews into futuristic shifts to get ahead of in the coming years.

Be sure to reference our GitHub repository containing all example code referenced in this book. If you have any questions or comments about the book or Cortex-M development in general, feel free to contact us directly. We have listed our email addresses here for your convenience.

- Zach Lasiuk – `lasiukza@bu.edu`
- Pareena Verma – `pareenaverma@gmail.com`
- Jason Andrews – `jasona.1842+N1@gmail.com`

Finally, thank you for reading! We (Jason, Pareena, and Zach) are all passionate about helping Cortex-M developers change the world through technology. We hope that after reading this book, you are better equipped to create the next generation of smart, secure, and connected devices.

Index

`Packt.com`

Subscribe to our online digital library for full access to over 7,000 books and videos, as well as industry leading tools to help you plan your personal development and advance your career. For more information, please visit our website.

Why subscribe?

- Spend less time learning and more time coding with practical eBooks and Videos from over 4,000 industry professionals

- Improve your learning with Skill Plans built especially for you

- Get a free eBook or video every month

- Fully searchable for easy access to vital information

- Copy and paste, print, and bookmark content

Did you know that Packt offers eBook versions of every book published, with PDF and ePub files available? You can upgrade to the eBook version at `packt.com` and as a print book customer, you are entitled to a discount on the eBook copy. Get in touch with us at `customercare@packtpub.com` for more details.

At `www.packt.com`, you can also read a collection of free technical articles, sign up for a range of free newsletters, and receive exclusive discounts and offers on Packt books and eBooks.

Other Books You May Enjoy

If you enjoyed this book, you may be interested in these other books by Packt:

Designing Production-Grade and Large-Scale IoT Solutions

Mohamed Abdelaziz

ISBN: 978-1-83882-925-4

- Understand the detailed anatomy of IoT solutions and explore their building blocks
- Explore IoT connectivity options and protocols used in designing IoT solutions
- Understand the value of IoT platforms in building IoT solutions
- Explore real-time operating systems used in microcontrollers
- Automate device administration tasks with IoT device management
- Master different architecture paradigms and decisions in IoT solutions
- Build and gain insights from IoT analytics solutions
- Get an overview of IoT solution operational excellence pillars

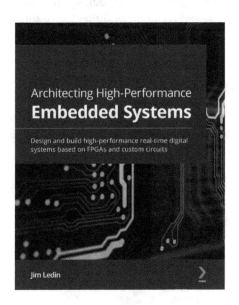

Architecting High-Performance Embedded Systems

Jim Ledin

ISBN: 978-1-78995-596-5

- Understand the fundamentals of real-time embedded systems and sensors
- Discover the capabilities of FPGAs and how to use FPGA development tools
- Learn the principles of digital circuit design and PCB layout with KiCad
- Construct high-speed circuit board prototypes at low cost
- Design and develop high-performance algorithms for FPGAs
- Develop robust, reliable, and efficient firmware in C
- Thoroughly test and debug embedded device hardware and firmware

Packt is searching for authors like you

If you're interested in becoming an author for Packt, please visit `authors.packtpub.com` and apply today. We have worked with thousands of developers and tech professionals, just like you, to help them share their insight with the global tech community. You can make a general application, apply for a specific hot topic that we are recruiting an author for, or submit your own idea.

Share Your Thoughts

Now you've finished *The Insider's Guide to Arm Cortex-M Development*, we'd love to hear your thoughts! Scan the QR code below to go straight to the Amazon review page for this book and share your feedback or leave a review on the site that you purchased it from.

`https://packt.link/r/1803231114`

Your review is important to us and the tech community and will help us make sure we're delivering excellent quality content.

Download a free PDF copy of this book

Thanks for purchasing this book!

Do you like to read on the go but are unable to carry your print books everywhere? Is your eBook purchase not compatible with the device of your choice?

Don't worry, now with every Packt book you get a DRM-free PDF version of that book at no cost.

Read anywhere, any place, on any device. Search, copy, and paste code from your favorite technical books directly into your application.

The perks don't stop there, you can get exclusive access to discounts, newsletters, and great free content in your inbox daily

Follow these simple steps to get the benefits:

1. Scan the QR code or visit the link below

https://packt.link/free-ebook/9781803231112

2. Submit your proof of purchase
3. That's it! We'll send your free PDF and other benefits to your email directly